D1301491

Fuzzy Rule Based Computer Design

John R. Newport

CRC Press
Boca Raton   New York   London   Tokyo

WINGATE UNIVERSITY LIBRARY

Library of Congress Cataloging-in-Publication Data

Newport, John R.
    Fuzzy rule based computer design / John R. Newport
      p.   cm.
    Includes bibliographical references and index.
    ISBN 0-8493-7834-6
    1. Computers—Design and construction. 2. Computer software—Development.
3. Computer integrated manufacturing systems. 4. Fuzzy systems. I. Title.
    TK7888.3.N493 1995
    621.39'2—dc20   95-23308
                                                 CIP

This book contains information obtained from authentic and highly regarded sources. Reprinted material is quoted with permission, and sources are indicated. A wide variety of references are listed. Reasonable efforts have been made to publish reliable data and information, but the author and the publisher cannot assume responsibility for the validity of all materials or for the consequences of their use.

Neither this book nor any part may be reproduced or transmitted in any form or by any means, electronic or mechanical, including photocopying, microfilming, and recording, or by any information storage or retrieval system, without prior permission in writing from the publisher.

CRC Press, Inc.'s consent does not extend to copying for general distribution, for promotion, for creating new works, or for resale. Specific permission must be obtained in writing from CRC Press for such copying.

Direct all inquiries to CRC Press, Inc., 2000 Corporate Blvd., N.W., Boca Raton, Florida 33431.

© 1996 by CRC Press, Inc.

No claim to original U.S. Government works
International Standard Book Number 0-8493-7834-6
Library of Congress Card Number 95-23308
Printed in the United States of America 1 2 3 4 5 6 7 8 9 0
Printed on acid-free paper

systems engineering series

series editor
A. Terry Bahill, University of Arizona

Preface

Design automation can decrease product cost, reduce development time, and simplify the transition to manufacturing. A highly automated process may also reduce development risk by allowing the engineering teams to complete test cycles earlier in the development and more rapidly than with traditional methods. The highly competitive microcircuit arena offers an example of a product area that has become automated from the design, prototyping and test, and manufacturing phases. In particular, much of the prototyping is attainable as high fidelity simulations that provide accurate behavior, electrical characteristics, and timings. Although a design team often fabricates "test chips" as a risk reduction measure, computer based *virtual prototyping* offers a seamless software environment between design and manufacturing using on-line design *synthesis*. As such, the processes and tools of microcircuit design and prototyping offer a metaphor for other product developments that rely on virtual manufacturing.

Embedded computer systems, used for real-time hardware control, are increasing in complexity and capability of their intrinsic networks, hardware architecture, and software design. System integration is the process of connecting, testing, and welding together the individual hardware and software components of such systems. Differences between hardware and software design methods that may have been tolerable with older, less complex systems, now render system integration an expensive, failure-prone enterprise. Despite such differences, some of these methods involve automated design. Consequently, the obvious leap to the new system integration capability is to select common hardware and software tools as the basis of a virtual prototyping process for computer systems. The microcircuit process serves as a metaphor and illustrates the need for a *system synthesis* capability to replace the fragmented simulation methods now used in system design.

The purpose of this text is to provide a vision of such a system synthesis capability. This capability rests on three separate technologies. One basis of this vision is common design and analysis methods for hardware and software. A strong candidate for a common method is the object oriented approach. This technique emerged from the software arena as an answer to user needs for increased reliability in increasingly complex designs. Systems engineers are now beginning to apply such methods to design, requirements capture, and engineering process controls. As such, object oriented design offers a common method for both hardware and software engineers. Another basis of the vision is artificial intelligence (AI) technology. Despite some rough starts, AI is now mature enough to constitute a high powered, low risk alternative to classical, algorithmic logic. Fuzzy logic can extend the capabilities of AI expert systems beyond the limits of binary logic. The third element is the decreasing cost and increasing performance of computer work stations. These computers offer the speed, high resolution graphics, and storage capacity to solve difficult engineering tasks at an affordable price.

The three elements constitute minimum needs for system synthesis, but they do not inherently provide this capability. Instead, system synthesis is an implementation that uses these elements. This text offers an overview of a possible system synthesis implementation and examples of its use. The topics and problems should be readily accessible to graduate students or engineering professionals.

A work of this magnitude can be successful only with the concerted efforts of numerous individuals. The author wishes to thank the many colleagues who offered suggestions about the manuscript. Especially, many thanks to Mr. Paul Williams of Alliant TechSystems for contributing the photograph of a processor module. The author thanks the hard-working staff at CRC Press, especially Mr. Robert Stern (Senior Editor) and Dr. Terry Bahill (Engineering Editor). Finally, many thanks to my family members for their encouragement during this project.

J.R.N.
June 1995

Contents

Part I

Introduction to design methods

The 1990s may show the same rapid evolution in computer design methods that occurred for automated hardware component design in the 1980s. These growth areas include object oriented analysis (OOA) and design (OOD) and virtual prototyping (VP). OOD offers the capability to manage the risk of a complex software or system development. VP allows a designer to use the process of simulation design, not just the final simulation products, as a model for system design using OOD.

The automated process of microcircuit design serves as a metaphor for the VP concept. A logic designer uses highly automated, graphical tools to capture the behavior, interfaces, and timing of the logic. This graphical description serves as an input to a code generator that can build a behavioral simulation model in a standard software language. The most common example of such a language is the very high speed integrated circuits (VHSIC) hardware descriptor language (VHDL) described in Section 8.3. The behavioral model provides a top level check of control and timing. A designer can then use a synthesis tool to build a physical simulation that accounts for each individual logic gate, signal pin, and physical layout. This simulation allows the designer to check electrical characteristics such as impedances, capacitances, and power dissipation. Another tool can generate a Netlist from the physical model, which is a series of instructions for the numerically controlled machines that generate the microcircuit in the foundry. At each step of this process, the quality assurance engineers can run the tests that are the basis of acceptance testing of the hardware in order to verify both the designs and simulations. A system tool set that can emulate this process of iterative design and testing from concept to prototype is an essential ingredient in the efficiency and practicality of VP. Such a capability is the topic of this book.

Behavioral simulation models are the nexus at which system, hardware module, component, and software designs meet. Even though engineering teams fabricate requirements and product baselines from behavioral simulations of these elements, the correlation of outputs from each stage is usually a manual process and therefore cumbersome. Artificial intelligence (AI) methods such as synthetic vision, natural language, fuzzy logic, expert systems, and virtual reality can integrate the various automated tools to provide a capability to quickly assemble and simulate an entire system design, as well as perform functional and performance trade studies. The AI methods allow designers of such environments to capture the experience of many system developers as synthetic rules, on which a system design assistant might be based.

There are many automated software design tools and numerous excellent texts on OOD. This text simply presents an overview of these areas to demonstrate their power and relevance to computer hardware design. However, this background illustrates a key system engineering process observation, that is, that many current computer designs, application softwares, and simulation design methods are independent and unrelated. Automated VP brings common methods to all three aspects. This can result in lower cost through the reuse of requirements and product designs. In addition, with the reuse of requirements, product designs, and common tools, more design teams can focus on each model. A more thorough review process can lower development risk by finding design errors earlier.

Just as a system is more than software, system OOD is more than software OOD. Instead, system OOD covers the design of both software and hardware elements. It is OOD and VP in conjunction that unify the software and system methods. Historically, differences in hardware and software design tools and methods made this approach unworkable. As Figure 1 shows, the newer design strategy involves three aspects of synergy between OOD and VP. These are common methods, system integration models, and reuse. The first concept involves an OOD approach to building the simulation software used for prototyping and risk reduction of the system. This step allows reuse of existing simulation software and requirements. Validation of requirements involves trade studies to demonstrate that the engineering details of the product description are technically correct. The validation can be quite expensive and time consuming, so a capability to use validated requirements from other projects can reduce risk and development time. The next aspect involves using the simulation software integration as a prototype of system integration. The final aspect relates to the reuse of previously validated requirements, functional baselines, and risk measures for the simulation as a test of the same reuse capability for the actual product.

The book is broken into three major parts. Individual chapters with references form the structure of each part. Exercises appear at the end of each part. These exercises may involve some research and design effort, so they

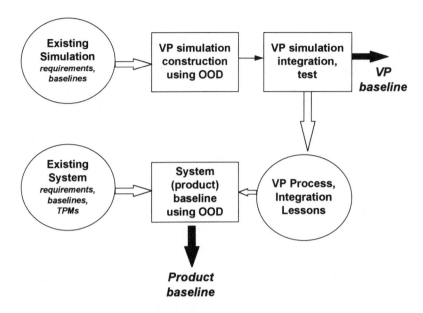

Figure 1   Relationship of OOD and VP in design.

are not necessarily geared for individual student efforts. As with any text-book exercise, the goal should be to learn the methods and process of design.

This part of the text presents design methods and key technologies of OOD and VP. Chapter one provides some introductory concepts and defini-tions. The second chapter presents models and management methods for the design process. The third chapter offers design methods of structured design (SD) and OOD. The fourth chapter discusses some automated tools for system design. The fifth and final chapter of this part presents VP and relevant AI technologies.

The second major part of this book contains system architecture funda-mentals and numerous, detailed examples. The architecture basics include the theoretical building blocks essential for insight into the examples. The examples represent a broad spectrum of applications, architecture strategies, and complexities. These examples also illustrate and detail the underlying rules and procedures for an automated system design method.

The third part of this text contains recommendations for further study and some historical trends. This includes some observations regarding gaps in the coverage that automated tools provide in terms of system VP.

Analytical methods are attempts to abstractly model the critical aspects of a problem domain. In classical physics, for example, the motion of bodies under very general assumptions can be described with high accuracy using the mathematical model developed by Isaac Newton. Newer domains of physics, such as quantum electrodynamics, use graphical methods to enable the analyst to visualize the problem.

Modeling methods in computer and software design target similar goals. The complexity of such systems must be decomposed into manageable pieces that can be modeled and visualized. Unfortunately, the complexity of modern systems threatens traditional simulation approaches to analysis.

The next generation system designer must have available powerful analytical tools such as those now available to a circuit designer. A microcircuit designer might build up a simulation of a new product from logic "libraries" that are extensively tested by a variety of projects. In addition, a hierarchical approach is available, starting from the most abstract behavioral models down to physical synthesis. The aim of this textbook is to demonstrate and teach how VP and OOA/OOD together can provide a similar capability for next generation system designers. These methods may offer maximum reuse of validated requirements, simulation models, performance results, and other expensive elements. This approach also allows the construction of the simulation to serve as a model for integration and testing of the actual system. Finally, this approach allows a designer the capability, not just to visualize the operation of the system, but also to interact with it using other human senses.

A widely cited story about the great James Watt involves how he was able to invent the steam engine. His answer was that he simply imagined in his mind the operation of the machine. If components suffered mechanical interference or some other problem, he visualized another shape to solve the problem, then reran the engine in his mind. Once such problems were solved, he built the machine. Modern computer systems are likely too complex even for James Watt to fully visualize. However, VP and OOA/OOD may offer design capabilities that allow even the average designer to surpass the accomplishments of such geniuses.

chapter one

Concepts, status, and goals

Embedded computer architectures have evolved through three separate integration methods during the period from approximately 1960 to 1990. The first-generation systems were *stand-alone*, involving little interaction with users except through an operator. The software complexity for such systems was inherently limited by slow execution speed and job turnaround. Therefore, processor and memory technologies became the most important technical subarea for system architects. The second-generation systems matured to become capable processor and memory systems that allowed the software to grow rapidly in executable complexity. Although direct user interaction became more feasible with such systems, the lack of routing capacity and the immaturity of interface standards made Input/Output (I/O) a technical subarea for which "guru" status was the price of admission. Thus the limiting factor for second-generation systems was I/O. The rapid evolution of hardware components and integration strategies brought the process full circle to the third generation. These technologies offer to existing software languages and environments much room for growth. Indeed, the performance of the software itself becomes a design issue with third generation systems.

In the transition from first- to second-generation systems, hardware complexity became an important consideration. Formal methods emerged from several large projects that permit more than simple monitoring of complexity, but allow it to be managed. In particular, structured analysis (SA) and SD became an important paradigm to ensure project success.

The transition to third-generation systems puts software in a similar situation. The complexity of many large software projects became unmanageable during the 1980s. However, experience taught that SA alone was inadequate for software, because the risk factors are so different in comparison to hardware design. In response to this "software crisis," a new method was developed, OOD. OOD has proven successful with second-generation systems. In order to fully exploit the risk driven approach, though, OOD must also be applied to hardware design. Once hardware and software use a common design methodology, a true risk driven approach to requirements analysis and testing is possible.

OOD is an emerging theoretical framework designed to emphasize system behavior instead of interfaces. Also, OOD emphasizes the iterative nature of system design instead of a top down, beginning to end approach. This feature of OOD makes it attractive to risk based design because it provides better communication and peer review among team members than a pure SA approach. Existing approaches focus on detailed requirements, whether documented in formal specifications or derived in some other manner.

In many ways, system design is approaching a crisis stage similar to that experienced by software designers in the 1970–1980s. The author believes current system design practices mirror many causes of the software crisis. The beginnings of this evolution for system engineering are contained in the recent MIL-STD-499B, Systems Engineering Standard. This document is not just for military use. Instead, it is becoming a general framework for process control within a complex development.

Advocates of MIL-STD-499B often say: *Designs should be risk driven, not specification driven.* This captures the practical experience that many requirements are waived in the end stages of development as trade study results become available, often with no apparent harmful effects. The converse happens with nearly equal regularity, in which aspects of the design that are overlooked initially, and therefore unspecified, often emerge as critical features. MIL-STD-499B provides the process framework to move to risk based design with its notion of technical performance measures (TPM). The standard requires that a project define and track the risk and quality metrics (i.e., TPM), but it does not define how to determine these metrics. Trial and error might produce a suitable set of TPM after many projects are completed. This approach, though, is quite inefficient.

The remark cited above from the advocates of MIL-STD-499B captures the essence of these deficiencies. Just as a new system engineering process must precede a risk based development, new *design methods* must emerge to produce repeatable behavior in systems engineering. OOD as a general framework for systems design appears to offer such a capability. The OOD method emerged from the software engineering realm and has proven to be a powerful method of requirements analysis and initial design capture. However, the application of this method to systems engineering is still theoretical. Therefore, it is important in subsequent discussions to retain the distinction between the two engineering realms.

Terminology can become quite confusing without some guidelines. The words "object" and "class" occur so commonly in discussions that some semantic rules are essential. In subsequent expositions, these words without adjectival modifiers are interpreted as common usage. For clarity in the technical presentations, the modifiers "software" or "system" are used, viz.: "system object," "software class." Common synonyms are also used later, such as "thing," "article," "item," and "others" for "object." The words "group" and "category" are common synonyms for "class."

In addition, OOD emerged from software engineering, which also defines OOA and object oriented programming (OOP). Although this is a textbook on design methods, the reader must understand that design represents only a fraction of the systems engineering protocol. For example, object oriented systems analysis (OOSA) might include system engineering trade studies. These studies are currently formulated on the basis of specification requirements that, in the subjective judgment of key technical persons, appear to be high risk. OOSA methods might more clearly identify, not simply risk areas, but also methods to measure and prove the risk of a key functional area or component. Object oriented systems development (OOSD) is the systems engineering equivalent to OOP, and is the method by which the system functional baseline emerges and undergoes development. Finally, OOD for systems bridges these two stages in terms of

- Detailed characterization of the desired behavior of the end system
- Formal methods which produce the system functional baseline from this characterization
- Capability evaluations, such as system stability, requirements reusability, and system engineering process impacts.

VP is an emerging technology that relies upon AI methods such as natural language, synthetic vision, and speech processing. The designer assembles these capabilities into a simulation that provides a realistic environment for both the user and the system. This allows the designer to simulate system integration and testing as the software units for the simulation are integrated. The complexity of these software units makes reuse and OOD a natural approach. This is the crossing point for OOD and VP—the integration, testing, and reuse of the simulation software serves as a model for the integration, testing, and reuse of hardware and software in the product.

Finally, a note of realism is essential. The transition to OOD must be evolutionary for several reasons. One factor is the large investment in SA training and tools that no established facility can afford to undertake at once. Another reason is relative immaturity of OOD, especially for hardware design. Nonetheless, experience teaches that a systems engineering process that is focused on the production of specifications may fail in predictable ways, since no organization can correctly specify every possible pathological combination of behaviors for subelements. The new process must strive to identify, quantify, and mitigate risk. OOD is an important method that serves this goal.

1.1 Key concepts and definitions

In spite of the broad introductory comments, readers may be assured that a finite scope of problems is the subject of this text, not the universe of all design methods. This chapter provides some key definitions and establishes the top level range of later topics. This is presented as a series of questions whose answers may establish at least an intuitive basis for upcoming presentations.

♦ What is a system?

It is hard to resist simply quoting the dictionary definition for system, but this approach offers little real insight in the current context. In this textbook, "system" is interchangeable with a "computer system" consisting of hardware and software. The scope of the discussion may imply a single board computer, a mainframe computer, or a computer network.

A system must be identifiable in terms of some key features that characterize the manner in which it interacts with other systems. First, the system has recognizable boundaries. For a personal computer, the case that houses the internal disk drives and computer boards is this boundary. For an automobile or other vehicle, the body, tires, and other application specific features form the boundary with other systems. Simply put, it must be possible to distinguish whether or not an element is contained within the system. Second, a system is comprised of interacting elements of various types. The personal computer consists of processor/motherboard assembly, power supply, fixed disk drive, graphics card, and multiple function card (serial I/O, parallel I/O, timer/clock). Each of these elements is recognizable as a different type, but routinely interacts with other elements of the system. The automobile is comprised of a larger set of interacting elements, such as engine, transmission, brakes, and many others. Finally, the system elements interact synergistically. That is, the interactions of elements produce behaviors that individual elements cannot perform.

Within a computer system, software is but one of many elements. Yet, more and more, software becomes the personality of the machine. Just as each person is more than a personality (i.e., outward physical and emotional behavior when interacting with others), the computer system is more than software performance or functionality. On the other hand, when people say that they "like" another, or that someone is "friendly" or "cheerful," these personality traits may dominate exchanges within a group of persons. In the same manner, software behavior in large part determines the successfulness of the computer system design to the end user.

This leads to the final point: system requirements. The user of a computer system does not intend to invest thousands of dollars in a computer system out of intellectual curiosity regarding its behavior. Instead, the computer is a tool that must provide some capability to the user. This capability may be

- Repetitive calculations (e.g., spreadsheet)
- Simulation of some complex event
- Document processing
- Graphics (scanning photographs, video capture, drawing)
- And many others.

The great accomplishment of personal computer and work station designers is that these systems can perform many functions for a low delivery price. Still, some system designs may emphasize one capability, such as processor speed which relates to repetitive calculations versus some other factor, such as highly reliable operation. As subsequent examples illustrate, there is usually not a unique set of component requirements for validated system requirements. The trade-offs that a designer uses in developing the lowest level requirements from the system requirements involves both art and formal methods. However, the lowest level requirements must be *traceable* back to a top level requirement.

◆   *What other design methods are used?*

The older system (and software) analysis method is SD. This approach became popular in the 1960s and 1970s as a consequence of large developmental programs, especially the Apollo effort. This approach is, in essence, the divide and conquer strategy to large projects. This method has been successful on many large projects. However, experience also teaches some shortcomings that other design approaches address.

SD emphasizes the need to

- Define the boundaries of each element, usually based upon system functions or subfunctions
- Define and control the information that passes across these boundaries.

In practice, many frustrations emerge with this approach. First, some implementations fail to account for the iterative nature of design. Somehow, requirements are never quite finished in real projects, which presents a problem in SD because the designer must capture the detailed functional and interface requirements as design entities. Another problem with this approach is its complexity for large projects. Structured software design is, fundamentally, decomposition of software algorithms into low level procedures that call one another to perform a system function.[1] Structured system design is similar because it is based on the functional decomposition of system requirements from which the baseline design emerges. The problem, though, is that the interfaces can be linked together in complex (even mysterious) manners. Anyone who doubts this should attend the next design review for an upgrade to an older aircraft and recommend a major change to an Interface Control Document. The howls are likely to subject the remonstrator's hearing to an on-the-job injury.

OOD and SD are related, just as are algebra and arithmetic. In arithmetic, there is no apparent relationship of the expression $2 + 3 = 5$ to the equation $3 + 4 = 7$. However, if this is expressed algebraically as $x + y = z$, then when $x = 2$, $y = 3$ the first expression is obtained, while $x = 3$, $y = 4$ provides the second. Yet, both results stem from the same algebraic statement. OOD, then, provides a more abstract framework that is based on the character of requirements, not on the detailed requirements themselves. This abstraction is the essence of the power of OOD, but it also explains the reluctance of some designers to embrace this method, just as elementary students initially feel more comfortable with arithmetic than algebra.

◆   *What are the advantages of OOD?*

OOD is a very important technology for several reasons. These include

- Impact on the ability of an organization to transition within the process maturity model
- Impact on reuse of designs, validated requirements, and TPMs

- Provisions for more natural system requirements changes which simplify and speed upgrades.

The process maturity model that emerges for system engineering and design is likely to mimic the capability maturity model (CMM) of the Software Engineering Institute. A later chapter provides details on this model, but at least one important feature emerges from this discussion. An organization that practices specification based design, common with SD, is unlikely to penetrate beyond level 2 of the 5 levels. Instead, a risk based approach is a key to the optimizing design team, which is a natural adjunct of OOD.

The concept of class, discussed later, is also a natural adjunct of OOD. The object-class relationship formalizes the "leveling" process of the SA/SD discussion. This is an important feature because it allows OOD users to more simply reuse code. Reuse is of practical significance because regardless of the source language, most software designers rely on examples based on previous coding efforts as much as they rely on formal design documentation. System designers follow a similar practice. Validating new requirements is very expensive due to the prototyping and testing needed to lower risk. Therefore, many system designers reuse validated requirements from previous designs. In an upgrade program, the system and computer subsystem requirements might not change at all, even though the computer hardware might bear no resemblance to the original design.

This discussion also emphasizes how OOD can simplify system upgrades. Too often, the obsolescence of a single microcircuit requires that a user throw everything away and start over. An OOD approach jump starts this process because it may permit designers to bypass much of the traditional design validation activities.

OOD is a powerful concept that helps computer designers become much more efficient. It is, in essence, a method to "work smarter, not harder" to achieve the same end result.

◆   *Who is developing the system OOD methods?*

OOD design is a very new and active area of systems engineering research. In addition to journals, magazine articles, and textbooks, professional organizations such as the National Council on Systems Engineering (NCOSE) are focal points for discussion of system OOD. In addition, there are usually many papers presented at aerospace and technology conferences such as National Aerospace and Electronics Conference (NAECON), Government Military Electronics Conference (GOMAC), Tri-Ada, and meetings of other standards groups such as the Society of Automotive Engineers (SAE).

◆   *What is VP and how does it relate to OOD?*

VP emerged from the AI arena called virtual reality. Virtual reality is a multimedia simulation that augments AI subareas such as natural language, synthetic vision, and speech processing. Natural language provides a programmer interface similar to spoken language instead of a computer language. This capability can be coupled with graphical methods to form a powerful design tool. Speech processing provides a computer system with the capability to speak and hear in a manner that mimics human interaction. Synthetic vision gives the computer the capability to see and process images. These subareas, and others, allow the computer system to simulate not just complex hardware systems, such as high performance aircraft, but also to interact with the human operator.

As simulations, prototypes, and actual products grow in capability and complexity, the boundaries between these elements blur. For example, some designers have fashioned a synthetic landing aid for transport aircraft that provides an artificial visual image in low visibility conditions, as well as a flight director that can perform the complete instrument approach and landing (including flare and application of brakes!). The flight simulators on which the pilots learn to operate this equipment, in many cases, contain the same computer equipment as an actual aircraft. At what point does the aircraft system end and the simulation system begin? This same technology is also now emerging in other industries, especially for automotive applications.

The simulation software for the virtual prototypes is complex enough that OOD may be the design approach. Therefore, the relationship between VP and OOD can be simple yet powerful:

> *Integration, testing, and reuse considerations for*
> *construction of the prototype serve as a model for the*
> *integration, testing, and appropriateness of reuse for the*
> *actual product.*

OOD can provide powerful insight into requirements validation, system integration, and reuse. Since reuse is an increasingly viable and valuable approach to reducing development cost and decreasing time to market, OOD and VP in combination may offer the maximum opportunity for the reuse of validated requirements, TPMs, and hardware and software elements.

◆   *What types of systems are the focus of this book?*

This text focuses on computer systems. There are many other types of systems to which the design methods apply, such as biological, economic, and social systems. To reduce the complexity and scope of the presentation, the topics avoid references to non-computer systems. Nonetheless, many of the problem solving methods are general.

Even the topic of computer systems presents such a broad target as to make complete coverage impossible. Although the theory and methods might be generally applicable, a specific problem domain helps to focus the discussion. The examples in Part II of this text involve embedded, real-time computers. Such systems are especially challenging to design due to the crucial timing and fault tolerance constraints associated with hardware control, as well as severe electronic packaging and environmental requirements. Examples include automotive control systems, military avionics, and numerous network systems for automated teller machines, medical "on-line" records, and others. This class of problems is important because similar projects exhibit the "systems engineering crisis" symptoms just defined, in which a timely and cost effective design delivery becomes exceptionally difficult. This stems from:

- High system complexity
- Difficulty in characterizing desired system behavior early in the design cycle
- Complex interactions among process control, maturity of the design activity, and the design method.

Other examples, some more theoretical in nature, provide additional insight into the methodology. For example, the simple feedback system described later is very general, yet is simple enough to design without automated tools. More complex examples often give the appearance of a design machine, into which requirements are fed, the handle is cranked, then out pops a design. The examples in this book cover a spectrum of complexity designed both to illustrate the design principles, yet also to demonstrate the power of automated design.

1.2 Status of system OOD

OOD, especially for systems, is a rather recent development. This presents a dilemma in selection of examples because the design tools are changing so quickly. However, a purely theoretical approach is not sufficient since it leaves the reader wondering:

> *How can one go down to the office/laboratory and start using OOD?*

This is an important question to answer, and the design examples presented later may offer insight. Still, it is counterproductive to focus on learning to use specific design tools. In fact, the software engineers responsible for OOD have repeatedly discovered that one of the fastest ways to achieve oblivion status for a project is to select a tool set before the metrics and process models are firmly in place. This is captured by the pithy warning that:

> *A fool with a tool is still a fool![2]*

Therefore, the correct answer to this question is that it is often impossible to begin system OOD at once. A mature design process, consistent with total quality management (TQM), MIL-STD-499B, and OOD principles, is essential to success with OOD. Since process control is largely a management function, not an engineering task, OOD can succeed only with the help of both managers and engineers. Therefore, for those interested in systems OOD, talk to the program manager, not the engineering technical lead, regarding how to start.

OOD as an engineering discipline is not isolated from other policy and technology trends in the computer industry. Controversy within the software community regarding source language (Ada or C++?) is likely to be mirrored within the systems engineering community. After all, automated design tools represent a significant investment for an engineering manager in terms of capital outlays, training, and job experience. Therefore, many managers are likely to insist that systems OOD is possible (even easy) with older automated tools. This may, in fact, be correct since Ada (and other languages) is not optimized for software OOD, yet many projects successfully use Ada based OOD. The other major industry trend is the attempt by various federal departments, including the Department of Defense, Department of Commerce, National Aeronautics and Astronautics Administration, to move to a common computer acquisition policy and technology base that avoids point solutions in order to formulate a larger, more cost effective industry support base. Therefore, systems engineers must expect more pressure to "use commercial" for such applications. Resistance to this pressure, when appropriate, cannot be presented in engineering terms dealing with computer performance. Instead, this policy must be rated on the basis of its risk in accomplishing desired system behavior, which is the essence of TPM and OOD.

1.3   Presentation overview

MIL-STD-499B relies on the classic life cycle model in which designs emerge from *concept exploration*, engineers refine and prove the design baseline in *demonstration and validation*, which then leads to *engineering and manufacturing development*. Subsequent discussions in this book rely on a similar model in which the three phases are design, prototyping, and manufacturing. This presentation culminates in rules and examples for expert systems. Chapter 9 focuses on design methods and Chapter 10 presents prototyping. Automated manufacturing is not a topic of this text, but the discussion provides summaries of key concepts as needed.

The structure of this text consists of three parts. The first part deals with the theoretical framework of system design and VP. A second part follows which contains numerous design rules and examples. Exercises appear at the end of each chapter. A final part provides a summary of design rules and trends.

Figure 1-1 shows the relationship of the various elements within the text. The process models provide the environment in which design must occur. The models might be most important to program management, but they obviously impact engineering costs, schedules, and design approaches. Such models include those of MIL-STD-499B. The design methods and tools must be selected for each individual project. Such topics appear later in the text.

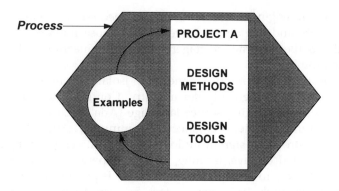

Figure 1-1   Presentation overview.

However, historical examples provide guidance in the selection of tools. Examples also provide starting points and valuable experience for each new project.

The first part of the text contains process models, design methods, and graphical tools. Process management elements include MIL-STD-499B systems engineering process, the CMM, concurrent engineering, and total quality management (TQM). The second topic involves principles of SD and OOD. The next topic of this part addresses the automated graphical tools available for system design. The final topic presents an overview of AI and VP technologies.

The second part discusses both formal models and design examples. Formal models include aspects of simulation, including VP, system communication models, and formal information security models. These lay the groundwork for the examples, which cover numerous applications from a diversity of industries, such as banking and financial systems, medical imaging and record keeping, aircraft embedded computers, and others. To keep the discussion tractable, only certain facets of these examples are fully detailed. Despite the informal treatment, the examples illustrate the formal methods, as well as other important features such as requirements and TPM reuse. The exercises contain additional examples.

The final part summarizes the impact of the design approach and provides a historical perspective. This also addresses some changes in policy related to funding of research and development.

References and notes

1. Booch, Grady. *Object Oriented Design, With Applications* (Benjamin/Cummings, Redwood City, CA, 1991), p. 18.
2. Believed to have been coined by Dr. J. Lake of the Defense Systems Management College. Dr. Lake is the focal point for the emergence of MIL-STD-499B. His university and continuing education systems engineering course are highly regarded.

Projects

Goals: There are no specific projects for this introduction. However, subsequent chapters provide projects designed to clarify and cement concepts. These are truly "projects," not exercises due to the labor involved. In addition, many correct approaches are possible. Each chapter also includes a brief statement of the goals for each project set.

chapter two

Process management

The system design process involves engineering and management efforts that deal with both top level requirements and those for technical specialty areas. MIL-STD-499B provides a formal definition of the systems engineering process for complex systems. The CMM emerged from work done by researchers at the Software Engineering Institute (SEI) at Carnegie-Mellon University. Concurrent engineering and total quality management (TQM) are engineering management methods that are dramatically altering the everyday work environment of design engineers.

System design is one of many elements for which systems engineers are responsible. Nonetheless, design may be the key element in which experience must blend with innovation. Candidate baselines, the products of design, can then be subjected to analysis. System analysis is not the focus of this text, although the results of such work are implicit in the examples. These and other tasks are the purview of systems engineering. MIL-STD-499B defines the relationships among these activities.

Another aspect of design is the capability to measure risk and predict product quality. These issues are critical to ensure that the average or inexperienced engineer can contribute to design evolution, not just senior, highly experienced experts. The CMM is a template by which design organizations can measure their capability in these areas. The capability to measure both quality and risk are essential tenets of the TQM movement.

The design process and capability are the cornerstones of selecting a design method. Program management must understand and communicate a commitment to these principles in order to assure a smoothly functioning organization.

2.1 MIL-STD-499B

The current systems engineering methods formalize the lessons learned from large aerospace developments beginning in the 1960s. There are many standardization activities related to this topic, such as MIL-STD-499B, an IEEE systems engineering standard, and the SAE Modular Avionics Working Group. However, none of these efforts are inconsistent with or provide more detail than MIL-STD-499B. For this reason, the Military Standard for Systems Engineering, MIL-STD-499B, offers a touchstone for discussions of formal methods.

MIL-STD-499B provides a general description of the systems engineering process for complex systems. The four areas that this standard identifies as the elements of the design process are

- Requirements analysis
- Functional analysis and allocation
- Synthesis
- System analysis and control.

This chapter explores some of the general features that MIL-STD-499B, as well as other system engineering documents, describe as accepted engineering practice.

At a sufficiently abstract level, each system design involves similar operations. The two competing design approaches are structured and object oriented, but this distinction is really more a matter of emphasis. This chapter provides some background on the design process. This overview, together with a few additional observations, provides the background necessary for an introduction to object oriented system design.

◆   *Requirements analysis*

The design process begins with a technical characterization of the system and its intended behavior in terms of functionality, performance, and interfaces. Concise, unambiguous statements of these requirements are critical so that both the customer and the designer understand the intent of the system. In addition, the requirements statements must be technically precise and complete so that the team can formulate acceptance tests that demonstrate the adequacy of the product.

Generally, systems engineering methods are valuable in the study of any complex system, but this discussion focuses on the design of computer systems. The definition of computer system in the first chapter cites the three distinguishing characteristics of recognizable boundaries, interacting elements, and synergism. Since this definition includes both hardware and software, further elaboration provides a firmer basis on which subsequent examples may rest.

Boundaries establish the breadth of the system. In more formal terms, both open and closed systems are possible. The analyst could mentally draw a ring around a closed system through which no data or control information flows, yet the system continues to change states internally and remains stable. An open system requires interaction with other systems. Regardless, the boundaries relate to the point where the data and control information for the system application must pass to the user or environment.

The interacting elements are commonly called *system functional elements*. However, to name something is not necessarily to understand it, so this begs the question:

◆ *What is a system functional element?*

Unfortunately, there is no universal answer, but some general characteristics are

- Well-defined boundaries, especially with respect to other functions
- A set of states characterized by recognizable, stable behavior
- A set of control actions.

Intuitively, a system functional element is a subsystem or subelement of the larger system. Therefore, the previous definition of system also describes a subsystem. The distinction between system and subsystem is usually clear from the context, but always involves a matter of scope and detail. In addition, a subsystem may not necessarily function in isolation, but a system must.

The control actions, often inputs from other functions, move the system function from one stable state to another. Later discussions show that it is not too difficult to relate this to the definition of an object. With some additional refinement, the linkage point between system and software terminology is that a *system functional element is an object; the system function is the behavior of the object.*[1]

The synergy requirement means, in simplest terms, that a collection of functions can accomplish a wider range of behaviors and control operations than the summation of the primitive operations of the individual functions. This leads, once again, to another widely used term, *application*:

◆ *What is a system application?*

An application refers to the top level operational characteristics that the system must provide. Therefore, a system application relates directly to operational and top level system requirements, but the designer must derive functional requirements (at least for complex systems).

A well-formulated system must have these characteristics, but this is an end point, not a design process. Many different paths enable a designer to reach this target from a clearly defined starting point. These include SA, OOD, or combinations. Regardless of the method, the designer must clearly understand the user requirements (starting point) and system behavior (ending point.) These features are often difficult to distinguish from the conceptual system implementation.

The system specification defines requirements for interfaces, performance, and behavior of the computer product. The requirements exist prior to, and are distinct from, existing or proposed implementations. For example, the requirement for a vehicle might involve transportation of two people at highway speeds, but an implementation is a red sports car. This distinction can be very subtle in practice, and is the source of much friction and misunderstanding between designers and customers. Requirements traceability offers a resolution to this conflict since the design must not contain lower level requirements that are not essential for the implementation of a top level requirement.

An implementation is a product baseline that includes hardware and software detailed requirements, functional partitioning, and component selections. Detailed requirements include interfaces, timing, performance, and similar features. The functional partitioning involves decomposing a system application into its elements. Component selections identify specific technologies and parts as building blocks for system functions.

Functional partitioning is an important design task that can impact the success of the remaining steps in the process. The definition of these functions and their interrelationships are somewhat left to the discretion and creativity of the designer. As an example, if one characterizes a complex biological system, some of the functions might be sight, sound, and touch, plus many others. A system level application might be night mobility. One can imagine an unfamiliar, dimly lit room that such a being must traverse. Arms are outstretched to search for obstacles, the eyes and ears are straining for clues. Even though each function is clearly recognizable and independent (i.e., the person might be blindfolded, but sound and touch still operate), no single function is likely to consistently provide success. Instead, a network manager, the brain, must integrate the sensor inputs and formulate a solution. The distinction, therefore, between the system and its implementation in this case is that the system is capable of night mobility, *NOT* that the system shall use sight, hearing, and sound in dim light.

Specification writing is a difficult art. It is important for a customer to capture years of lessons that are learned at a dreadful price, yet allow the designer the freedom to explore new approaches. A common complaint, about military specifications in particular, is that this balance is lost and the requirements dictate a single approach. This is clearly undesirable from both the perspective of both the customer and designer, but how can one not throw away experience? The answer lies in the exhortation quoted in the first chapter, *system design must be risk driven, not specification driven*. That is, the goal of the development process is not to draft industrial strength specifications, but rather to understand clearly the intended behavior of the system, as well as lessons learned from similar projects. The process control must involve risk analysis, not just documentation. The popularity of OOD is, in important ways, tied directly to these goals.

♦ *Functional analysis and allocation*

The design activity must decompose the system into functional areas. Each of the functional areas must meet lower level, "detail" requirements so that the system, as an entity, meets the top level requirements. Therefore, the functional analysis and allocation activities involve

- Assignment of top level system requirements to functional elements
- Establishment of the performance and testability associated with each functional (lower tier) requirement
- Careful definition of the relationships among functions.

In software terminology, this step is equivalent to unit design and testing. A designer must fully characterize the building block (system functional element or software unit) in terms of behavior and interfaces. Functional element testing is an important precursor to successful transition to the next phase.

One advantage of software OOD is that it more naturally enables the reuse of software units. The same comment is true regarding system functional elements. The advantage of OOD in system functional element design is that the requirements can be reused for other systems. Since requirements analysis typically represents a major portion of the design cost, this is an important advantage. The examples address this topic in more detail.

◆ *Synthesis*

Synthesis is the process of formulating a system configuration based on the functional partitioning. This includes evaluation of alternatives and verification that the selected candidate meets the requirements. This includes

- System element alternatives
- System element integration and verification
- Synthesis products (primarily the product baseline)
- Design completeness
- Interface compatibility.

The first component recognizes that early in the design process, many different system configurations may be possible. The designer must cull these alternatives on the basis of top level requirements. In practical terms, available manpower, cost, and schedule may limit the number of alternatives to five at project inception, then three alternatives one year into the project, and finally, a single approach after two years. The design baseline is the output from this step.

Many later discussions refer to this key portion of the systems engineering process with the description of *design synthesis*. In such contexts, design synthesis refers to the use of highly automated methods to convert input requirements into a product baseline. Microcircuit design and others serve as a process metaphor for design synthesis.

Synthesis is the functional element integration process whose output is a product baseline. This step is the proverbial proof of the pudding since functional elements are designed to predict behavior as part of functional analysis and allocation. If such predictions prove invalid, it is necessary to return to functional analysis. Therefore, synthesis and functional element analysis are inherently iterative. The design completeness and interface compatibility are the activities during which the predicted behavior is tested.

◆ *System analysis and control*

System analysis and control is the management of requirements and baseline configurations. This involves trade studies, risk analysis, and process control.

The trade studies are the analysis and tests by which system alternatives and component selections emerge. Typical selection criteria are maturity, cost, and performance.

Risk analysis involves definition and monitoring of TPMs. Once again, OOD offers advantages in this aspect since it increases the ability of design teams to reuse TPMs from one project to another. Unfortunately, many program managers permit key technical people to define the TPM uniquely for each project. These definitions are rarely carried from one project to the next except in very general, vague terms. Also, the process of monitoring a TPM helps to formalize the design iteration that is an essential feature of OOD.

Finally, process control involves scheduling, reviewing, and approving these activities among the various design teams. A key modern management concept is the *integrated product development team* (IPDT). The IPDT method forces designers from each technical subarea to distribute design issues for team review on a regular basis. In addition, efficiency usually demands common design practices, such as standard development and documentation tools. This approach simplifies the burdensome job of a program manager because it produces outputs in a common format that are more easily evaluated.

◆ A design process example

An example of the development process may cement these concepts. Examples later in this book do not cover every aspect just discussed, but focus on specific features that illustrate key points. Therefore, a simple, inclusive example is relevant that addresses the evolution of requirements and the design.

A trap into which many fall prey is to select an automated design tool, then select a design method compatible with the tool. This has the effect of allowing the tool designer to select the overall design process for the project. Sometimes, the choice of an automated tool may be based primarily on available funding. This procedure allows the selection of a support tool to drive overall project cost, schedule, and method. This should not suggest that automated tools are worthless. Indeed, complex system design can be nearly intractable without such tools. Instead, the system engineering process must relate to specific relationships among team members based on acquisition policy, the technology available, as well as the engineering roles of customer and contractor.

Figure 2-1 shows the stages involved in the production of a system design. The diagram includes development focus, some products that formally define the baseline, and narrative text that further explains each step. The amount of feedback and feed forward among these steps is somewhat controversial, but the diagram includes certain loop backs for emphasis.

The process begins with a need statement, or, for government customers, a mission need statement. The need statement contains a top level customer description of the product. As an example, suppose one is shopping for a

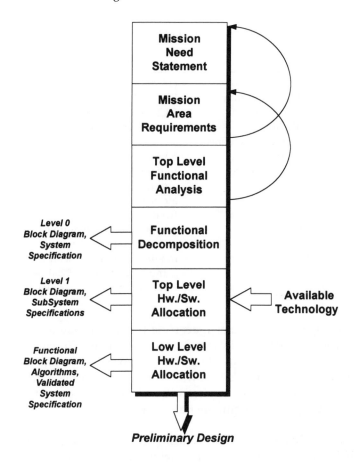

Figure 2-1 System design process flow.

new car. The following features might constitute the need statement for the new vehicle:

The mission shall consist of a daily commute consisting of 5 city miles and 20 highway miles each way; a carpool one week per month consisting of 15 miles of city driving twice daily; and an annual vacation consisting of 800 miles of highway driving. The vehicle shall safely, comfortably, and affordably transport two adults and two children for these missions.

The mission area requirements analysis involves decomposing this need statement into segments. The first priority stated in the need statement is safety; comfort is the second priority, and affordability is third. However, there obviously must be some trade-offs among these requirements. The analyst must decompose these top level requirements into mission phases. For example, the safety equipment for highway driving may be different for highway trips than for city driving due to the difference in speed. Comfort is another consideration since, for example, there may be more ambient noise

Table 2-1   Mission Area Requirements for the Auto Purchase

Mission area	Requirement
1. Mission scenario	(14,750 miles per year)
1.1 Daily commute	81.4% of total mission miles
1.2 Car pool	13.2% of total mission miles
1.3 Vacation	5.4% of total mission miles
2. General requirements	
2.1 Vehicle life	5 years or 75,000 miles
2.2 Safety features	remote door locks
	anti-lock brakes
	air bags/belts
	front wheel drive
2.3 Comfort	air conditioning required,
	cruise control
2.4 Cost	monthly payments, with trade
	must be less than $250 per
	month with a 5 year loan

at highway speed, yet extensive acoustic treatment may add weight that detracts from affordability in terms of fuel economy. To continue the example, each mission phase must be broken into segments and assigned top level performance, as Table 2-1 shows. Although the entries in this table tend to identify equipment, for example, "anti-lock brakes," a correct mission area statement must provide requirements, such as:

The vehicle shall provide positive braking action on all highway surfaces. Positive braking is achieved when the vehicle delivers minimal likelihood of wheel lock, skids, or uncommanded directional changes.

The next topic is functional analysis. This involves definition of the top level system functions needed for the mission area requirements. Examples include accelerate, cruise, and stop.

Each of these must be correlated with the mission phases. For example, during the highway phase of the daily commute, the accelerate phase involves a highway on-ramp culminating in a merge to highway speed. Performance is not really a factor at this point, except that the vehicle must be at highway speed from a standing start within a normal length on-ramp, say one half mile. At this point, the system designer must be able to provide a verbal description of the capabilities necessary for the system to meet top level requirements.

Functional decomposition breaks these top level functions into subfunctions. The cruise segment, for example, might require numerous capabilities to meet the safety, comfort, and cost criteria:

- Constant speed control
- Cabin temperature control

- Engine control and monitoring
- System diagnostics.

The same subfunctions might be included for many top level functions, but the decomposition is important because it provides a map for low level performance requirements.

The top level hardware and software allocation is the initial definition of the relationship of available technology to functional areas. If no technology is available, the functional decomposition must be readdressed. Perhaps the most important feature of the process in Figure 2-1 is that available technology feeds into allocation, not earlier. A system designer must be aware of technology, but technology must never be allowed to drive mission needs or functional requirements. On the other hand, a need statement occasionally cannot be met with available technology, but it is important to have a clear understanding of the mission and functional requirements before this judgment is made.

The low level hardware and software decomposition involves preliminary design within each subfunctional area. Performance and interface requirements stem from this process. A completed design involves iteration of the last three steps, as well as a test cycle.

SA emphasizes a top down design approach. The focus is on starting at the beginning in Figure 2-1 and delivering the design after a test cycle. OOD emphasizes the feedback among these steps. The object oriented approach allows the decision to proceed to be based on a risk assessment instead of completion of documentation, such as specifications. Also, the object oriented approach allows the design quality to be measured at each step, not just during the final test cycle. In practice, most system designers use a combination of both object oriented and structured elements. However, automated tools, as previously suggested, can heavily influence this approach. The discussion continues with an overview of each design approach.

2.2 CMM

The impetus for the CMM has its origins in the widely touted software crisis of the 1980s. The root cause of this crisis was that application software began to grow in complexity without commensurate sophistication in analytical capabilities. It began to seem that no person could possibly understand all aspects of such software. Many looked to the emerging TQM movement to find solutions for these problems. Two important aspects of TQM, despite the diversity of approaches, are the capability to measure both risk and quality.[2]

However, lest system designers begin to make smug comments about "those software types," one might consider the technical difficulty of integrating such software applications into increasingly complex hardware environments. A careful study of the CMM reveals that it provides a development process template for any complex item, either hardware or software. Consequently, the CMM has received much attention in recent years as a systems engineering model to implement the MIL-STD-499B process requirements.[3,4]

The structure of the CMM involves five different maturity levels:

1. Initial
2. Repeatable
3. Defined
4. Managed
5. Optimizing

The initial level occurs when no formal process definition is available to a new program, which forces the team for each new program to develop its own process. This can be a complicated undertaking that few program managers are likely to schedule or fund. Unfortunately, a common default answer for this problem is to purchase an automated, graphical, computer based tool for the project. This, in effect, allows the tool designer to define the process for the program. Sometimes the match is good and the tools contribute to team success. Other times, the tools do not match the level of complexity or functionality of the deliverable product. In such cases, the tools can actually inhibit development. If a project of this type fails, often the program manager is unable to discern the reason for such failure, short of the need for more dollars and time. Therefore, program success depends entirely on the work habits of the team members. At this level, the loss of a key team member may doom the project.

Program managers within a level two organization can rely upon standard management tools for cost, schedule, and product performance analysis. This means that historical data are available to indicate which products were delivered within cost and schedule. However, this is still a high risk process because product performance is often unverifiable until integration is complete. Even worse, this means that historical data may not be useful to a new program since it may not be clear why a particular program failed or succeeded. Such organizations might require little else other than standard reporting formats, such as the cost, schedule, and status reporting (CSSR) used widely in the Department of Defense. The documentation of product performance focuses on the system specifications and the associated test methods. The danger of this approach is that complex systems can be developed and pass through this process, yet fail at the test phase. This is a specification based process, not a risk based approach. An important principle of risk management is to identify failures as early as possible in the development.

A level three organization fully defines both management and engineering activities. However, each program manager may have different priorities in terms of data collection. That is, the team may understand the process, but the quality and risk data are unlikely to be available to other program managers.

A level four organization accumulates performance data in a standard format for each program. This allows program managers to identify any high risk activities during the program planning phase.

The level five organization optimizes its performance by modifying its procedures in response to the historical data. To avoid chaos and reversion back to level one, this fine tuning may not be done for individual programs, but must be the result of periodic, top level management decisions. Another feature of such organizations is *continuous improvement*, another TQM concept. However, it is quite challenging to build an organization that can respond to constant change.

This background may suggest that a systems engineering organization that complies fully with MIL-STD-499B must be a level five organization. It takes only a little thought to realize that this is not correct. Answers to the following questions offer insight into this paradox.

- What are the primary deliverables from a design activity?
- Is the design process documented and are team members trained in the process?
- Does the organization have a consistent definition of TPMs for all programs?
- Does the organization actually track TPM performance for all projects?

Process definition is a management tool that helps to eliminate wasted engineering effort so that the organization becomes more competitive. A simple feedback system of control theory offers some final concerns. The feedback from this system can be convergent or divergent. The divergent behavior is caused either by underdamping, insufficient responsiveness to the sensor input, or overdamping, excessive sensitivity to the sensor input. The program director, or other second level manager responsible for program managers, must achieve this process balance to sustain a level five capability.

2.3 Concurrent engineering

The development of a complex system tends to follow a progression of activities (see Figure 2-2). A well-planned transition between these phases is critical to program success. Many older engineering process management approaches treated each phase as distinct, with rigidly cast inputs and outputs. Unfortunately, experience demonstrates that there is significant overlap and interaction among the various phases. *Concurrent engineering* is a modern management technique that accounts for this interaction.[3]

Although some practitioners may dispute the boundaries among the engineering phases shown in Figure 2-2, one observation is universal. That is, the later in the development that the team uncovers a problem, the more expensive the problem is to fix. One rule of thumb is that the development cost increases by a factor of 10 for each phase. Therefore, a component selection in the design phase may represent a cost of $100. However, if the selection is erroneous and remains undetected until system test, the design team must pay $1,000,000 to correct the problem. Although this example exaggerates reality, it does point out that early decisions can have disastrous

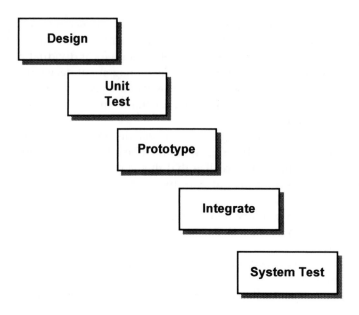

Figure 2-2   Engineering phases of product development.

consequences. Some more detailed examples may illustrate the pitfalls more clearly.

A designer carefully researches available data processors and selects a high performance, vector processor that meets all top level requirements. However, during unit testing, the engineering team discovers that it is impossible to write test software because there are no external test points that provide visibility into the internal operation of the logic. This problem must be solved with a custom integrated circuit that performs parity checking and other hardware checks for all of the input or output signal lines to this processor. Trudging onward to prototyping, the fabricators discover that the processor package is unconventional and requires a very expensive socket for assembly. Also, the processor is very large, so some circuit card layout changes are necessary to simplify fabrication. Eventually, several prototype circuit cards emerge from the production facility for the system integrators. Unfortunately, the system integrators discover that power dissipation from the processor causes other circuits to overheat and malfunction. This requires the addition of a custom cooling system. Finally, system testing begins in order to demonstrate the engineering features of the product to the customer before delivery. At this point, the customer decides that the unit is much too heavy and expensive, the reliability appears to be poor, and the unit is difficult to repair. The team must face the difficult choice of significant redesign or abandonment of the project. If engineers associated with each phase had been involved in the review of the product baseline, it is likely that some or all of these problems would have been discovered during the design phase.

The concurrent engineering approach to this problem is to address the dependencies among engineering phases by involving the entire team in all phases of the effort. This is, above all, a peer review process. The functional baseline that emerges from the design phase must undergo a thorough analysis by those who are responsible for other phases (for example, system testing) before the transition to unit testing formally occurs. At this point, the designers become part of the review process and those responsible for the fabrication of test elements begin their primary activities. This process minimizes the likelihood that significant problems will remain undetected until the late phases.

This approach may seem inefficient from a manager's perspective. After all, the activities involve a much larger number of people so that the development cost must be higher. This is valid only for products that pass through a conventional engineering process with no undetected problems. As this is quite unlikely for complex systems, the concurrent engineering approach is actually more cost effective.

One management method has become a popular approach to the implementation of concurrent engineering. The *integrated product development team* has two distinctive characteristics. First, the team is organized around the delivery of a single, complex product. Second, engineers responsible for each phase of the product development remain on the team through the entire engineering cycle.

To revisit some problems that engineering experience reveals, many organizations require engineers to support a product only when its development activity coincides with their specialty. For example, a logic designer might be reassigned to a new effort once unit testing on the first product begins. This approach allows management to organize the engineering staff around technical specialty areas, which is clear and easy to track. Unfortunately, this approach tends to amplify the cost of fixing solutions found late in the design process because new engineers must learn, diagnose, and correct the problem.

The classic organizational structure is shown in Figure 2-3 for the A to Z Corporation. There are two major problems with this approach. First, it is unlikely that the engineering directors can formulate training plans and personnel policy in sufficient detail and generality to meet the needs of each program manager. This forces each engineer to undergo extensive on-the-job training each time a new program requires the services of that engineer. The other problem is that the working level engineer is distant from the inputs of the customer who provides the source of funding. This makes requirements formulation needlessly complex and promotes miscommunication. However, the strong point of this approach is the clear lines of authority and responsibility.

A newer structure that an *integrated product team* (IPT) might adopt is shown in Figure 2-4. This sort of organization is based on program managers directly interacting with customers, as before. However, it also relies upon directors responsible for each engineering program phase instead of engineering specialty area. This structure addresses the two problem areas of the other organization, but introduces another problem. If someone shows you Figure 2-4 on your first day at the A to Z Corporation, you may have

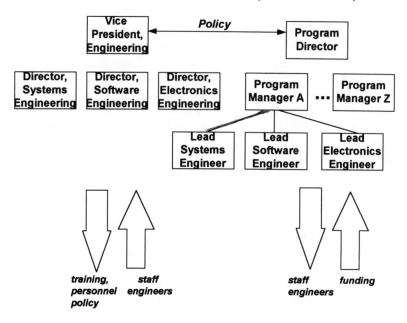

Figure 2-3   Classic engineering organization.

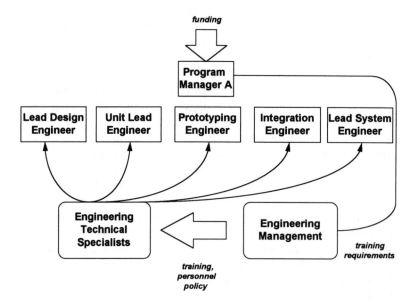

Figure 2-4   IPT team structure.

difficulty identifying your "boss" on this chart. In this type of organization, individuals that are willing to assume responsibility and take the initiative in projects are most highly prized by the various managers. This illustrates another complaint of senior specialists, though, who complain of significant responsibility with little authority.

The IPT concept meshes well with both topics of this book, OOD and VP. OOD inherently is based on the interactive design process, which is also the key to the operation of an IPT. Also, prototyping may be a separate engineering phase with a specialty manager for this activity. Therefore, VP neatly fits into the IPT team structure.

There are no magic bullets for management of new programs. The IPT concept is the most likely approach for complex systems, so engineers on such teams must be willing to accept responsibility for engineering success. The reward can be a greater sense of accomplishment when any team member is successful. As mentioned before, risk management and constant change are the operatives for next generation program managers. Engineers must learn to thrive in these and subsequent development environments.

2.4   TQM

The theoretical basis of TQM is statistical process control. This is a branch of mathematics that provides the underpinnings of measurement, observability, and predictability in a system for which statistics offers a valid model. TQM itself is a management approach that provides rules for implementing statistical process control. These rules are sufficiently general so that they apply to more than the conventional quality engineering tasks, but also provide insight into the design process.

The theory of stochastic processes provides the statistical underpinnings of TQM. An undergraduate course in statistics describes the Gaussian (normal) distribution, which is the basis of the familiar "bell curve" approach to sampling. The importance of the Gaussian distribution lies in the Law of Large Numbers, which states that the statistics of any system become Gaussian for a sufficiently large number of measurements. This trend is surprisingly rapid. In the case of a Uniform distribution, which assumes that all numbers within some range are equally likely to occur as samples, three iterations of measurements produce a recognizable Gaussian distribution.[4] Two parameters, the mean and standard distribution, fully describe the behavior of a Gaussian distribution. If these parameters become time dependent, a stochastic process is the result. This theory was used by Einstein in his classic description of the Brownian motion of gas particles. Mathematicians also use the more entertaining example of an exceptionally intoxicated person looking for car keys under a street lamp. Deming's classic text offers many powerful examples in product manufacturing.[5]

Despite the theoretical basis for TQM models, managers need not be mathematicians to apply it. The TQM approaches teach managers how to identify aspects of the process that must be measured. The next step is to accumulate data for these elements over enough time so that a statistically significant number of measurements are produced. These measurements then offer insight into the strong and weak areas of the production process. This approach couples with the human nature of the workers, who ordinarily will strive to output high quality products if they are given the training and independence. The team as an entity (production worker, engineers, and managers) can develop solutions to improve the weak areas. This approach is called *continuous improvement* in TQM, and may offer benefits such as decreased production cost, shorter time to market, and higher quality.

An example may clarify these ideas. Consider service at a fast food restaurant. The service goal is to allow each patron to traverse the line and reach the service counter in fewer than 5 minutes. As the manager collects statistics, she notices that one server is consistently slower than this goal. Rather than berate the employee, the manager asks the employee for an explanation. Initially, this person is unable to provide an answer, but offers to monitor more closely and report back later. Later in the day, the answer becomes apparent. Each day, a large number of individuals arrive with very complex special orders. These persons have a favorite server, namely the person in question. However, this only partially answers the issue, since it does not address why the special orders are so time consuming. The answer becomes clear by observing the cook's methods of preparing these orders. The solution is to either modify the menu or simplify the preparation instructions. Since the customers consistently request a small number of items, the latter solution is obvious. This simple example illustrates the three top level steps of TQM. These are measure, analyze, and improve.

OOD and VP fit naturally into the TQM approach because they are based on recognition of the iterative nature of design. However, each engineer is likely to be challenged to identify risk areas of the design process and accumulate measurements in the form of TPMs. This is an evolving area of applied TQM and unfortunately, engineers must rely principally on intuition since historical examples are not prevalent. However, engineers should begin to accumulate such information without being asked, because building a TPM, which is a mathematical model of a risk element, requires the time and patience of many individuals. These characteristics are often in short supply in the midst of a design crisis.

2.5 Process impacts on design

All newer approaches stress team involvement in every aspect of the process. This requires much more interaction and peer review than earlier methods. Junior engineers may complain of endless meetings in such processes, while senior engineers resent the apparent lack of design independence. This frustration is often a lack of recognition that perceived design progress may occur in jumps with this approach, not a steady slide towards completion. The team approach also is the basis of the system or software *factory* model, which is characterized by the capability to sustain quality and production rate despite the replacement of key team members.

MIL-STD-499B and the CMM are specific models based on concurrent engineering and TQM. These models identify and quantify design risk as the primary program cost and schedule drivers. OOD and VP are engineering methods by which designers can respond to such impetus.

OOD supports these management concepts because of its more efficient nature in comparison to other methods. OOD may offer a greater opportunity for the reuse of validated requirements, functional designs, and TPMs. This capability can decrease development time and cost.

VP also supports these approaches because it offers early insight into system functionality and performance. This can identify design, fabrication, or production problems of complex systems long before a working prototype is available. Therefore, VP represents a risk reduction tool. In addition, if the VP software design uses OOD, the software integration can serve as a model for the actual system integration.

A diversity of book topics share a common theme. Career planning guides, financial planners, and quality assurance texts all emphasize that the ability to react to change is an essential ingredient to success. Nowhere is this observation more true than with engineering process methods.

References and notes

1. Booch, Grady. *Objected Oriented Design, With Applications* (Benjamin/ Cummings, Redwood City, CA, 1991), p. 514.
2. Paulk, Mark et al. Capability Maturity Model for Software, Version 1.1. CMU/ SEI-93-TR-24, Software Engineering Institute.
3. Torino, Jon. *Managing Concurrent Engineering* (Van Nostrand Reinhold, New York, 1992), Chapter 2.
4. Maybeck, Peter. *Kalman Filtering* (Addison-Wesley, New York, 1989), Chapter 3.
5. Deming, W. Edwards. *Out of the Crisis* (Massachusetts Institute of Technology, 1982).

Projects

Goals: These projects help the reader become familiar with MIL-STD-499B and its related documentation. The second project also provides some preparation for the next chapter.

1. Obtain a copy of MIL-STD-499B. This can be ordered from U.S. Naval Publications in Philadelphia, or through your technical library. Consult the Department of Defense Index of Specification and Standards (DoDISS) for the most current version.

    (a) Review the publications list of standards groups such as SAE and IEEE. Can you find any process standards for systems engineering that are similar in content to MIL-STD-499B? Compare and contrast the content of these standards with MIL-STD-499B. (Hint: Start with IEEE-P1220.)

    (b) Review the definitions in MIL-STD-499B. Notice that these definitions relate to the engineering process, not necessarily to the design. Which definitions do not apply to a design baseline, but only to the engineering process?

    (c) Review figure 3 of MIL-STD-499B, which shows the process inputs and outputs. How do these inputs and outputs contribute to or relate to risk assessment?

2. Obtain product literature for automated, system design tool vendors. Does the literature describe OOD usage? Does the literature describe use of this tool for SA? Compare and contrast the examples given.
3. Obtain detailed information on the CMM, from SEI, if necessary.
   (a) Are there conflicts in the requirements and MIL-STD-499B and the CMM? Give examples.
   (b) Is an organization that complies with MIL-STD-499B assured of attaining level 5 maturity? Why or why not? Give examples.
   (c) How does the CMM relate to risk assessment for the program manager of a particular product? How does the CMM impact an engineering manager of a specialty area such as a software or systems design group?
   (d) Does the CMM assume a matrix management approach, a line management approach, or does it support either? Provide examples.
4. Pick some computer technology area, such as data processors, signal processors, software, or others.
   (a) What are some risk factors that a program manager might face for this technology area?
   (b) Formulate a risk matrix that provides an overall ranking for this technology. Assign a numerical score that relates to overall priority of this risk factor.
   (c) Fill in the risk matrix for your technology area assuming a highly fault tolerant, embedded, real-time system for which a working prototype must be available in three years. Is your technology high or low risk? Why? What are some activities that could be started to lower risk?
5. Explain the IPDT approach and provide examples.
   (a) Compare the IPDT approach with the classic acquisition method of "black box" organizational structure. In this program structure, written requests for information must pass from customer engineering to customer management, to prime contractor management, to subcontractor management, then to subcontractor engineering. The answer must traverse the reverse path. Engineering issues are addressed formally at periodic reviews and coordination meetings. What advantages does this traditional approach offer? What are the disadvantages? (Hints: How much time does each method require, how many people are involved, who is responsible for engineering decisions, and how does the customer recover lost money if a poor decision is made?)
   (b) In comparison to the traditional approach, what advantages does the IPDT method offer a program manager? What are the disadvantages?

chapter three

Design methods

The last chapter provided an overview of evolving management techniques. These methods require that someone, often a design engineer, identify and quantify development risk. The manager can then concentrate available funds and personnel on the highest risk areas. The engineering basis of the risk analysis may not be apparent, though. One aspect of such analysis is formal, repeatable design methods.

At least two design approaches are common. These are SD and OOD. While purists may debate the merits of each with nearly religious fervor, practicing engineers usually adapt elements of both. The selection may be based on customer preferences, available automated tools, and personal experience. This chapter examines these two methods in detail.

For clarity, the distinction between analysis and design must be revisited. MIL-STD-499B defines analysis as requirements definition and validation, as well as definition of final product tests associated with particular requirements. Design in MIL-STD-499B refers to formulation of proposed solutions for requirements. This is comparable with conventional software usage for which analysis involves identification of requirements and design relates to synthesis of solutions for the requirements. This textbook involves the latter, or product formulation. Nonetheless, a clear understanding of the requirements must precede design.

Much of the debate regarding the two methods focuses on the top down design philosophy that is the cornerstone of SD. This approach is, in essence, a divide and conquer strategy in which a complex problem is decomposed into related simpler problems. The primary distinction between SD and OOD is the tactics of divide and conquer.

The decomposition of SD is based on data flow in the system.
The decomposition of OOD is based on the behavior of
components.

In some cases, therefore, SD is more efficient, but for other systems OOD produces better candidate designs. Simulation design and VP, in particular, are well suited to the OOD method. In addition, OOD for VP allows deep insight into possible strategies and problems of actual product integration. The remainder of this part provides the theoretical basis of these ideas, while the next part provides examples and design rules.

Despite these advantages, SA/SD will continue to be a factor because

- SA/SD represents one of the historical origins of OOA/OOD and many older systems are documented as SA/SD.
- Real world design teams will continue to rely on a combination of methods best suited to the project.
- Graphical methods and automated tools for OOD are still evolving. The chapter on OOA/OOD revisits many of these same issues.

Therefore, subsequent discussions should not be considered disparagement of SA/SD despite the prevalence of OOD concepts in these discussions.

3.1   General concepts

One important observation is that designers often prefer graphical approaches in order to visualize the problem. One might hand a logic diagram to a design team and some hours later find the group formulating a baseline. On the other hand, if one provides the team with a lengthy, complicated specification, it may be difficult to even force individuals to read the document. Even the senior designers may read the document at home or on an airplane, where sufficient privacy prevails to the extent that it allows concentration. Reading, after all, does not give the appearance or personal sense of accomplishing "real" work.

This observation is surprisingly general. People who program signal processing for sensor interfaces may use the AT&T Signal Processing Flow Diagrams. Even computer initiates seem to prefer a windowing environment to operating systems with command line entry. SA/SD and OOA/OOD (and other methods) have diagrams for each discipline.

The concept of a *state transition diagram* is common to both approaches and has its roots in control theory. Mathematical representations are from matrix theory, especially linear algebra. A modeler can describe a system in terms of a *state vector*, which is an $N \times 1$ matrix, in which the equations that describe system behavior have N independent variables. The dynamics of the system can be described with an $N \times N$ matrix that transforms a state vector. The most common examples are linear systems, in which the system equations are of the form:

$$\frac{\overrightarrow{dx}}{dt} = \underset{\sim}{F}\overrightarrow{x}, \text{ with } \overrightarrow{x} \text{ the N x 1 state vector}$$

The concept of stability is a key ingredient of all control theory. In the simple differential equation above, the solution is exponential, with divergence or convergence determined by the sign of the components of the $N \times N$ coefficient matrix.

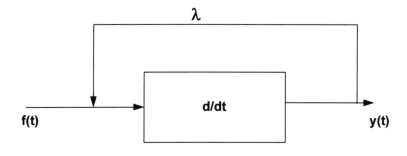

$$y = f(t) - \lambda df/dt$$

Figure 3-1   Simple feedback transform.

Another extension of control theory is the *transform centered design*.[1] The mathematical concept of a transform, such as a Laplace transform, is a black box function that accepts certain inputs and produces other outputs. In control theory, the outputs might be scaled and fed back as inputs that represents a *feedback loop*. Control theory provides the tools to identify whether such configurations are stable. Figure 3-1 shows an example of a feedback system for the derivative operator. From elementary differential equations, for the temporally stable solution of $y = 0$ (a homogeneous equation), then the solution is exponential and the output is bounded if $\lambda < 0$ and unbounded if $\lambda > 0$. If λ can be imaginary, then oscillatory solutions are possible. To further generalize, engineering analysis of control systems often involves Laplace transforms. If factored polynomials occur in the denominators of the transforms (this happens frequently), the solutions for the transform can be calculated using Cauchy's theorem. The roots of the polynomial are called *poles* because the denominator becomes zero for these values. Whether the roots are real or imaginary, and on the positive or negative axis, identifies whether the resulting exponential solution for the transform is stable. Much of the control theory analysis involves finding the poles and identifying the quadrant in which they occur to evaluate stability. Therefore, graphing the poles gives a visual indication of the stability of the system.

The state transition diagram captures much of this formalism. Each system has certain states that are stable, unless acted upon by an external agent. A modeler can characterize these states in terms of the operations that the system performs, such as

- Initializing
- Idling
- Fetching inputs
- Calculating
- Dumping outputs.

Notice that the labels of these operations are verbs. Each state has events that cause the system to enter or exit the state. For example, the system exits the "calculating" state when the results are available or an error occurs. Recovery from the error might cause the system to enter either the initializing or idling state. The system enters the "dumping outputs" state when new outputs are ready after the calculations are finished. Transitions between states are the consequence of discrete (usually not continuous) events.

Another approach involves modeling the data flow within a system. A *data flow diagram* consists of connected inputs, operations, and outputs. The input is either provided externally, such as initialized data or sensor values, or is the result of some intermediate operation. An operation is simply any system behavior that changes the data. The output is the result of this change. As a very simple example, consider the integer multiplication binary operation. Two integers are inputs and the operation provides a single output. Notice that these operations are different than control operations and transitions. Integer addition requires a new data flow diagram, but probably shares an identical state transition diagram with integer multiplication. This is the essence of the SA/SD formalism.

OOA/OOD have a series of diagrams that capture a functional baseline. One key element is the object diagram (OD), which characterizes each system object. The OD supplements the state transition diagram (STD) with additional information about the behavior of each object.

In a complex system, there may be many levels of these diagrams. For example, the "fetching inputs" state above may involve reads from a global memory, several different sensor inputs, or message reception and processing at the system network bus. Therefore, a state transition diagram for the five states above is a *level zero diagram*, which is the most abstract. The inputs processing is a level one diagram, but even this might consist of multiple subdiagrams. A similar decomposition of the data flow diagrams is often needed as well. The purpose of this *leveling* of the diagrams is to make system operation more comprehensible by providing a hierarchy.

Figure 3-2 demonstrates a top level comparison of SA/SD and OOA/OOD. Once again, the distinction is data structure and control vs. control messages and components.

Finally, an algorithm is often represented in beginning design classes as a *flow diagram*. In this approach, predefined symbols represent key calculations and decision points. This method is unsuitable for complex systems, though, for many reasons. First, it quickly becomes cluttered and incomprehensible because it contains both control information and data flow. In essence, a flow diagram includes both the state transition and data flow diagrams. Also, the data flow diagram often implicitly defines a hardware architecture solution because of the sequencing of operations and data flow. It is very important to provide system designers a design method that is independent of the implementation to provide the most general solutions.

Once again, these issues are to be revisited in later chapters. The important ideas, though, are that it is possible to describe the system design process in very general terms for which graphical methods are suitable. In the SA/

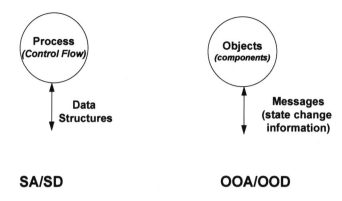

Figure 3-2 High level comparison of the design methods.

SD approach, the system states ("verbs"), and data flows are key design features.

3.2 Structured analysis and design

SA and SD originated from several large projects of the 1950–1970 era, especially space projects such as Apollo. It is no coincidence that during this same era, computer technology matured very rapidly. Computer aided design grew side by side with emerging microcircuit and software technologies since it became obvious that many repetitive design procedures could be automated to allow faster cycle times. These exciting times were the environment for emergence of modern systems and software engineering. SA/SD was one of many tools developed in this era.

Earliest graphical methods, such as programming or system flow charts, were useful in organizing and understanding complex systems. However, flow charts also have serious problems as an analysis tool. The problems stem from the mix of information types on these diagrams, such as data flow, control, timing, behavior, and other aspects. In addition, a flow diagram incorporates the physical design since it brings in hardware elements such as disk drives or other storage elements. Formal analysis must be based on a logical representation in graphical or mathematical form, not a physical form, if the model is to be valid for any design implementation.

Perhaps the main contribution of the SA/SD approach is a model for system development and use of graphical problem solving methods. In very general terms, a system can be described in terms of a series of actions, or "verbs" in later OOD terminology. The relationships of these actions can be sequenced in time or may be the result of some other event. The temporal relationships may involve continuous action or discrete events.[2]

The SA/SD method captures system behavior primarily with two graphs, the state transition (discussed in the introduction and later) and data flow diagrams (DFD). The rules for building these graphs are addressed in the examples. However, a verbal description helps to cement the top level design approach.

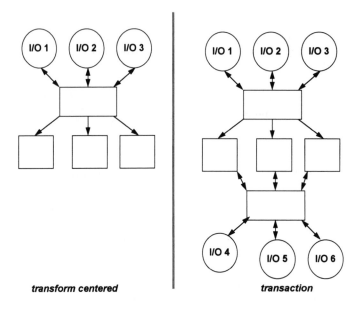

transform centered   transaction

Figure 3-3   Examples of structures for analysis.

The DFD is an essential element of the SA/SD method. This diagram is built up from processing (represented as circles) and data flows (indicated by arrows) as Figure 3-2 demonstrates. The difference between a DFD and a software flow chart is primarily that a DFD is a logical representation while a flow diagram is physical in terms of storage devices (disk drives, tapes, nonvolatile memory), processing elements, and interfaces.[3] The concept of building a logical representation is the basis of any analytical method.

Structured analysis, as the name might suggest, is based on the structure of the logical model of the system.[4] Two different SA methods are *transform centered* and *transaction analysis*. Figure 3-3 illustrates the structures most suitable for each type of analysis. Transform centered analysis is more commonly used and assumes that the I/O elements occur predominately at the top level of the logical structure of the system while calculations and decision logic are at the lowest structural levels. Transaction centered analysis is based on a model in which the decisions and calculations are at the center of the structure.

Several structural concepts provide the basis of the taxonomy used in SA. The three types of elements are afferent, efferent, or transform. An afferent element involves data flow from the outside up and into the top level structure of the logical model of the system. Efferent data flow is the opposite, from the top of the logical structure down into subelements. Transform elements involve modifications to data, such as calculations or decision logic. Elements can exhibit multiple data element types, although the goal of SA is to separate these behaviors as much as possible.

Once a logical model is constructed, the width (number of top level elements) and depth (number of subelements) must be balanced. Some related concepts are fan out, also called the span of control, which indicates the direct dependencies among elements. Fan in is the number of immediate superordinates. The scope of effect is the number of related or dependent modules.

The SA approach involves constructing the DFD, which is a graphical, logical model of the system. There are many methods of building such a model, but there are some general design rules. First, the model should maximize the fan in. Second, the fan out should not exceed 10 elements. Third, the model should minimize the number of pathological I/O operations, which are represented in the graph as horizontal data transfers instead of vertical transfers. Finally, the scope of effect should be a subset of scope of control. These rules produce a highly modular design in which each element can be considered a "black box" so that interface definition and control becomes a primary strategy of successful system integration.

In system design, analysis based on communication requirements has a long and successful history. Many designs of the 1980s are based on the federated architecture approach that treats each computing node as a "black box" that must be integrated into a larger system using standard communication interfaces. In addition, the communication equipment for such designs often represents the most difficult performance requirements for which to match hardware. The data driven methods, and DFD in particular, are important methods for such systems.

Newer system designs may be much more highly integrated. In these cases, the processor and memory systems may represent the performance bottlenecks. Therefore, DFD may not be the optimal approach for the design of such systems.

3.3   Object oriented design

OOD may be the most popular phrase in both the system and software design worlds since "I'm from the government and I'm here to help." In the midst of such enthusiasm, it can sometimes be difficult to gauge the importance of new technologies such as OOD. However, there is no doubt that OOD is emerging as the preeminent design method for new systems and software.

Still, this technology is very immature, especially as it relates to systems engineering. On complex programs it is better to walk, not run, so OOD may not be appropriate for near term, aggressive programs.

The terminology itself becomes a touchstone for practitioners. There is little use in fighting this phenomena; if one wishes to engage in OOD, one must speak the language. This section provides the underpinnings for discussion of OOD, such as

- Objects
- Classes
- Polymorphism
- Inheritance

- Abstraction
- Encapsulation.

Ada is widely cited (correctly) as a language that does not intrinsically provide support for OOD and object oriented programming. Therefore, the following discussions of these topics as they relate to software includes both Ada and C++ examples.

The following discussions include many examples of state transition diagrams. The approach is intentionally intuitive at this point and will be firmed later. However, the state transition diagram is such a key to system development that early exposure is valuable.

3.3.1   Objects

An object is the smallest design entity, the atom of a software or system chemistry set. As the first step in OOD is identification of the objects, a careful definition is essential to application of the method. An object has[5]:

- *State*
- *Behavior*
- *Identity.*

State refers to both the static and dynamic properties of the object. A static property is simply initialized, but a dynamic property can change. Behavior refers to state changes induced by message passing between objects. Finally, identity is that unique set of properties that distinguish an object.

Consider a very simple example, a digital signal line. The three possible states are on, off, or disconnected (see Figure 3-4). Some external agent induces the state transitions. For example, the signal line might be a discrete attached to a push button, connected through an optical isolation circuit for noise immunity. When the button is depressed, the signal line state becomes on. When the button is released, the state is off. When the electrical power is removed from the optical isolation circuit, the signal line is disconnected. The identity of the signal line may be related to:

- Its physical location in a backplane
- The name of the switch to which it is connected
- The name of the register into which it places its output
- Other system dependent factors.

Consider a balloon. One can fili the balloon with varying amounts of air, stretch it, or otherwise change the shape of the balloon. The shape is not what allows one to recognize the object as a balloon. On the other hand, consider an airplane. The shape is critical to the performance of the aircraft. Indeed, if one stretches or bends an airplane, the owner might be unhappy. In this case, the shape is an important characteristic of the airplane. However, the

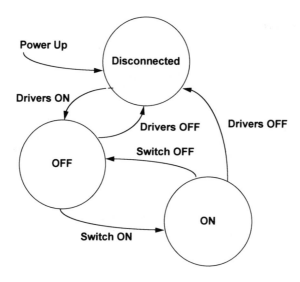

Figure 3-4 State transition diagram for a digital signal line.

precise shape depends upon the aircraft type. A Cessna 340 is quite different in appearance from a Boeing 727, yet both are aircraft.

A simple, but more detailed example, may help to provide further insight. As the next topic describes, a class refers to a set of objects that share common characteristics. This concept can be made very formal by appealing to the mathematics associated with set theory. For example, the ordinal numbers are a class with digits such as 1 or 2 as objects. In a more intuitive fashion, classes in system OOD are equivalent to the system functions that emerge from decomposition.

Therefore, consider all of the navigation equipment in an aircraft. The identifying characteristics of such equipment are that it measures a signal related to aircraft position, compares it with a crew selected nominal value, and then produces an error output that displays the corrective action that the crew should undertake. Common object instances are tactical area navigation (TACAN), long range area navigation (LORAN), global positioning system (GPS) area navigation, self-contained navigation, and many others. Figure 3-5 shows an STD for navigation elements.

A TACAN system measures the line of sight distance between the aircraft and the ground radio station that the crew selects. The position information is encoded with a phase that allows the crew to measure the direction to or from the station. The crew provides a nominal input based on navigation charts and the clearance. The LORAN and GPS systems are functionally similar. However, LORAN uses two widely separated radio stations and GPS uses satellite transmitters.

Self-contained navigation is the inertial reference system and digital computer that continuously computes the aircraft position. The crew initializes the software and sensor systems before flight with a precise determination of the starting position. The system then updates this position based on

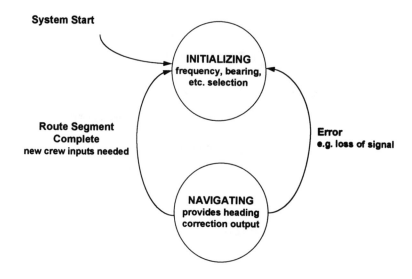

Figure 3-5   Navigation system transition diagram.

"self-contained" inertial sensors that do not rely on external radio stations. Instead, these sensors directly measure aircraft acceleration and angular velocity, which the digital computer integrates to produce a position solution.

The navigation elements prove to be valuable examples that can illuminate many issues. Subsequent discussions revisit these topics.

An object, as a design entity, has certain behaviors that uniquely distinguish it from other objects. Object states and state transitions neatly summarize these behaviors. However, objects also usually have other characteristics that do not serve to identify the object. In terms of software objects, data flow in and out of an object does not characterize the object.

3.3.2   Classes

A class is an object template. That is, the definition of a class provides the most abstract definition of a spectrum of objects with identical interfaces and behaviors. As Booch defines it

A class is a set of objects with a common structure.[6]

This concept is often easier to understand based on some examples instead of the abstract definition. One of the examples of this textbook is a distributed network, such as a telephone exchange. A simulation of the connections within the exchange might include "user" as a class. All users have some common characteristics, such as the capability to receive or place calls or the capability to handle only a single call at any time. However, there might be many different types of users on the system whose communication requirements differ, such as telephone users, on line computer users, or facsimile machines. There might be 1000 objects of the telephone type, 100 of the computer type, or 500 of the facsimile type. The general behavior and interface operation of each of these users is identical. The only difference in this example is data rate.

A class is the most general characterization of a system behavior. Top level system functions might be captured in the logical model as a class. For example, all area navigation elements in an avionic system might accept sensor inputs, perform some processing, and provide a time, distance, heading, and ground speed to the next waypoint. As another example, consider a computer network complex. There may be many types of computers that communicate within this network, but the interfaces and general behavior of each node are likely to be identical. Therefore, "computing node" might be a class in this case, with objects such as work station, personal computer, or minicomputer.

The concept of classes and objects is very important to the issue of reuse of software, validated system requirements, TPMs, and other design elements. Subsequent examples offer some insight into these issues.

3.3.3 Operations and polymorphism

Operations are the actions that objects within a class can take. In a system description, the nouns of a sentence might describe the objects and the verbs might identify the operations. As the last topic describes, a system designer can assemble those objects with a common structure into a class. However, a particular type of operation may span many classes. Polymorphism refers to the characteristic of such operators that the class context defines the precise meaning of the operation.[7]

Two common compiler usages of polymorphism are *operator overloading* and *function overloading*. Operator overloading occurs frequently in both Ada and C++ programs. For example, consider overloading of the arithmetic operators shown in Figure 3-6. The implementation details are not critical to this discussion, but rather that the user can extend the conventional arithmetic operators to operate on other types of variables. Function (or procedure, or other structural entities) overloading is similar. Function overloading in C++ requires that the compiler evaluate context sensitive function calls, such as those shown in Figure 3-7. In this figure, the compiler determines which function to use based on the data type of the parameters passed in by the calling program. The compiler evaluates this at compilation time; it is not hard coded by the user.[8]

Polymorphism is quite useful in software design because it renders the design analysis independent of data types, as well as other I/O features. Now recall the earlier comments about changes to interface control documents in a system upgrade. The data types must not determine the behavior of the system objects.

Polymorphism is further detailed in later design examples, but a quick illustration is probably helpful. Consider the navigation objects described in chapter one. One of the operations on these objects is to display aircraft

```
Package BOOLEAN_DATA is
      begin
function "+"
      (LEFT , RIGHT, Boolean) return Boolean;
function "-"
      (LEFT , RIGHT, Boolean) return Boolean;
function "*"
      (LEFT , RIGHT, Boolean) return Boolean;
function "/"
      (LEFT , RIGHT, Boolean) return Boolean;

end BOOLEAN_DATA;


-------------------------------------------------

with BOOLEAN_DATA;      -- boolean operator definitions

procedure TEST is

      X, Y, Z          : Boolean := True;
      A, B, C          : float ;

      begin

            C := A + B ;                          -- floating point add

            X := False ;
            Y := False ;
            if (.....) then X := True ;
            if (.....) then Y := True ;

            Z := X BOOLEAN_DATA.+ Y ;      -- Boolean add

end TEST;
```

Figure 3-6   Operator overloading.

```
void square (x : int , y: int) ;

            •
            •
            •
            •

void square (x : Boolean , y : Boolean) ;
```

Figure 3-7   Function overloading.

course error. However, this correction format depends upon which navigation mode the crew uses. For example, if navigating by VOR/THCAN receiv-
ers, the crew counts the number of dots of deflection and multiplies by 2.5
degrees per dot to calculate the amount off course. On the other hand, the
same procedures and indications apply for a localizer (or Instrument Landing System approach) *except that each dot represents 0.5 degrees*! Therefore,

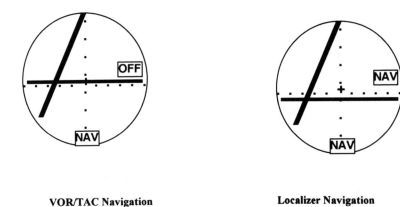

VOR/TAC Navigation   **Localizer Navigation**

Figure 3-8   Navigation system indicators.

calculation of the correct heading to get back on course depends upon which navigation object is in use (see Figure 3-8).

3.3.4   Inheritance

In some applications it is convenient to partition classes into elements called subclasses. This is equivalent to the mathematical set operation of dividing a set into subsets. Perhaps the most widely used examples are splitting the real numbers into integers, naturals (positive integers), or finite ranges of integers. Inheritance means that an operation defined for a class is also valid for a subclass.

The simplest example of inheritance from software is data subtyping. For example, the Ada fragment:

type UNITS is integer range 0...9

defines a new type that has all the properties of an integer, but a restricted range. This means that integer arithmetic operations are still valid with the subtype; they are inherited from the parent type integer.

The fact that the operator is valid on a subclass does not necessarily mean that it will produce correct results. One problem that this simple example illustrates is closure. Addition of 6 + 7 (both of type UNITS) produces 13 (not of type UNITS). This result would cause a run-time error involving constraint checks, but it would not generate a compile-time error because of the operator definition.

Another type of inheritance deals with attributes. For example, if a computer enclosure is characterized by its length, width, and height, all subtypes have these same characteristics. Therefore, a full ATR and 3/4 ATR unit enclosure both have length, width, and height, although the values of these variables differ.

In C++, the language permits inheritance directly in the class declarations.[9] The format of this declaration is, for example:

```
class alphabet {
     int length ;
     int script ;


.......


class English : public alphabet {


.......
```

in which the class English automatically inherits the attributes of length (number of letters) and script type.

Therefore, inheritance can be a useful property because it allows relationships among classes that define, in effect, super classes. This property can also lead to very confusing code design, so some caution is necessary in its use.

3.3.5   Abstraction

The notion of abstraction involves emphasis on those features that are important to an item. Therefore, abstraction is an important aspect of the description of an object, since, as with any modeling construction, one must avoid details that do not contribute to the fidelity of behavior of the conglomerate model.

Data abstraction is the simplest example from the software world. This involves user defined data types, such as the Ada fragment:

```
type MY_INTEGER is new INTEGER;
```

which then causes the user to respecify binary operations associated with MY_INTEGER, as in the discussion of polymorphism.

In fact, there are several types of abstraction of which data abstraction is only one.[10,11] The notion of abstraction, as just mentioned, involves capturing the important behavior while hiding the unimportant behavior, primarily implementation details. Consider a more complicated example. Suppose that you wish to write portable Ada code and therefore do not wish to include hardware details in the application code, such as register formats. On the other hand, it is likely that the communication management software must verify proper I/O transactions by examining low level registers, such as a status register. Although the size, bit pattern, and other characteristics of status registers for various I/O types are likely to be dissimilar, the general categories of errors might be similar. Therefore, it is possible to specify in a package a common interface:

```
type CURRENT_STATE is (IDLE , VIE , NORMAL_BLOCK_TRANSFER ,
               PRIORITY_BLOCK_TRANSFER);
```

```
type MESSAGE_PRIORITY is (NORMAL , HIGH) ;
```

```
type ERROR_CODE is (NORMAL , ERROR ) ;

type STATUS_REGISTER is record
        CURRENT_STATE       := IDLE ;
        MESSAGE_PRIORITY  := NORMAL ;
        ERROR_CODE            := NORMAL ;
end record;
```

These types of definitions appear in the package specification and represent the data structures that are visible to other packages. The implementation, though, is dependent upon the structure of the underlying hardware. For example, on some machines it may require accesses to multiple registers to fetch this information, while other machines may contain all the information within a single register. Still other machines may not have a hardware status at all, which may dictate that the Ada implementation update these data items in the software, based on commands that are passed into the communications software.

The package TEXT_IO in Ada provides a more complex example. This package provides standard functions to write to the user terminal screen or read from the keyboard. These functions include[12]:

PUT_LINE();   (write text to screen with a carriage return)
PUT();   (write with no carriage return)
GET_LINE();   (read text from keyboard or file assuming carriage return)
GET():   (read with no carriage return)
NEWLINE;   (carriage return).

The interface is the same for all Ada programs, but the implementation is machine dependent. An Ada application program simply includes a call to one of these functions, but a special library is linked during compilation to provide this service.

A more complicated example is a computer disk drive. The important characteristics to model are access time in milliseconds and storage capacity in megabytes, MB. Functions similar, or identical, to the TEXT_IO functions can be formulated to provide a file interface. The detailed operation of the hard disk controller and its special memory is irrelevant to the application program, so, as before, a special library can be linked in to couple with this standard Ada interface.

A third example demonstrates a virtual bus interface. A communication model establishes a user interface for which many types of implementations are derivable. However, the operations and data structures might be similar to the contents of Package TEXT_IO. This is often called *file structured I/O* even if the underlying hardware is a data bus, memory, disk drive, or other device. Therefore, the concept is extensible to memory structures (virtual memory), access software for disk drives (virtual drives), or other devices. Although standard software interfaces extract a performance penalty, this approach greatly improves the portability and reusability of the software. Often the performance penalty is quite small, so the trade-off is not difficult to make. Common virtual I/O communications interfaces, for example, require around 6% extra performance.[13]

The concept of abstraction is quite important to object oriented systems design. Later in this book, abstraction is the key to building a model of the core avionics system of an aircraft, in effect a *virtual avionics suite*. The model permits the designer to work abstractly with system requirements in early designs, instead of being limited by capabilities of some particular implementation.

3.3.6 Encapsulation

Encapsulation is the method a designer uses to accomplish *data hiding*. Data hiding is really quite a simple principle. As a general rule, only those variables and data structures necessary to modify the operation of a function or procedure should be visible outside that function or procedure. This avoids messy data consistency problems in software, which can be very difficult to diagnose.

Suppose that an application program contains a package that includes the specification for all of its global data. Invariably, variables get moved to the global section as patches for various problems over time. The danger of this approach is that complicated timing loops among procedures and functions can occur. Suppose, for example that the program contains two Ada tasks, one that performs I/O and the other that performs a calculation. The I/O task might complete, but it could also block while waiting for some hardware event, such as an error recovery. It is likely that the I/O feeds the calculations. If the I/O task blocks, it is suspended and the task involving calculations executes, which will now use old inputs for the calculations. The exception handler for the I/O eventually unblocks and this task starts again. By now, though, the damage is done. It is very easy in complex software designs to get exceptionally complicated timing relationships among units.

At the system level, encapsulation means that a system object must make visible to the remainder of the system only those features that are essential to communicate state or behavior information. Intermediate states, for example, which do not interact with external processes, should not be visible at the system level.

Return one final time to the signal line state transition diagram. Each state in the diagram interacts with the remainder of the system either through the optical drivers or the push button switch. An internal process should not be included on a system level diagram.

3.4   Summary and significance

In order to manage the complexity of large systems, developers can adopt a divide and conquer strategy in which discrete, low level elements are the responsibility of a small number of individuals. However, without some framework for the design process, integration of these elements into a system may be impossible.

SA/SD intrinsically formulates an abstract system model on the basis of the structure of the data and control flow in that system. The DFD is the key tool that enables an analyst to visualize this flow. This approach works very well on "black box" designs that involve loose coupling between computing nodes using standard interfaces.

OOA/OOD formulates the abstract model on the basis of the behavior of elements. Decomposition in OOA is based on concepts such as abstraction, encapsulation, and modularity. In addition, sequencing, such as concurrent execution, is a consideration. This approach is more abstract than SA/SD and therefore somewhat more difficult to visualize. The key graphical methods are class diagram, object diagram, and STD.

Highly integrated, parallel computer systems are emerging from design shops whose complexity is enormously larger than previous generation multiple computer systems. In order to maximize the quality of these systems, an iterative design process based on TQM principles is essential. One aspect of this is that testing of the design must be continuous, not a pass-fail check at the end of design. Prototyping has proven historically to be invaluable in risk reduction, but the time and cost to prototypes may be excessive for highly parallel systems. The designer needs tools similar to those of the integrated circuit designer who can start with behavioral models, synthesize gate level models, perform sophisticated physical simulation including analog effects, all prior to the delivery of the first test chips. VP is the technology that provides the capability. If the VP is developed using OOA/OOD, this development can also serve as a process model of system integration.

References and notes

1. Yourdon, Edward. *Managing Structured Techniques* (Yourdon Press, New York, 1979), p. 94.
2. Ward, P. and Mellnor, S. *Structured Development for Real-Time Systems* (Yourdon Press, New York, 1985), pp. 45 ff.
3. Yourdon, op. cit., p. 145.
4. Yourdon, Edward. *Structured Design* (Prentice Hall, Englewood Cliffs, NJ, 1979), pp. 112–187.
5. Booch, Grady. *Objected Oriented Design, With Applications* (Benjamin/ Cummings, Redwood City, CA, 1991), pp. 77–84.
6. Booch, op. cit., p. 93.
7. Rumbaugh, James et al. *Object Oriented Modeling and Design* (Prentice Hall, Englewood Cliffs, NJ, 1991), p. 25.
8. Schildt, Herbert. *Using Turbo C++* (Osborne McGraw-Hill, New York, 1990), pp. 469–472.
9. Ibid., pp. 473 ff.
10. Cohen, Norman. *Ada as a Second Language* (McGraw-Hill, New York, 1986), pp. 382–384.
11. Booch, op. cit., p. 42.
12. Cohen, op. cit., pp. 630 ff.
13. Historical data based on extensive testing by the author.

Projects

Goals: The following projects may help to overcome some of the differences in jargon among the various design methods. Also, this provides some practice in system decomposition. The intent is not to establish formally correct graphs, though, as this is the topic of the next chapter.

1. Identify and consider any complex system (e.g., a human, international corporation, aircraft.)
   (a) What are the functional areas (classes) for this system? Describe clearly the function that each performs. Specify a short name for each class
   (b) What are the identifying characteristics of each class?
   (c) Identify object instances within each class. That is, describe the specific members of each class. For example, if the class identifier is computer display, the objects might be monochrome cathode ray tube (CRT), color CRT, or liquid crystal display.
   (d) What are the operations for each class with respect to the defining characteristics? That is, specify the control or data manipulations of the identifying characteristics which are common to all the objects of a class.
   (e) Describe any relationships between classes in terms of natural language. For example, if receptionist and telephone are two classes for an office system, the relationship is (class) secretary answers (class) telephone. The relationship in this example is "answers" and in the jargon of OOD is called a *link*.
   (f) Make a chart in which each class is represented as a box. Include the name of the class in each box. List the characteristics of each class below its name. Illustrate and label relationships between classes with line segments connecting the relevant classes.
   (g) Congratulations, you have just constructed a class diagram. The next chapter provides more details.
2. Consider a complex system (the one from the previous project works just fine.)
   (a) Perform the functional decomposition of the system using the system engineering methods of Chapter 2.
   (b) Compare and contrast the functional elements with the classes of the first project. Explain any overlaps or differences.
   (c) Compare and contrast the concepts of OOD with the engineering process of the CMM. Are there any discrepancies? Does OOD lend itself to level 5, or does the design method have any impact?
3. Prepare a STD for a complex system. Once again, the system of the first project is helpful to consider since it allows comparisons with earlier results.
   (a) Identify the system states.
   (b) Identify the events that cause transitions among the states.
4. Prepare a DFD for a complex system, as the following list of questions describes.

    (a) Specify the top level system functions. This is equivalent to the functional element decomposition discussed previously.

    (b) What information transfers are required between these functional elements?

5. In your perusal of help wanted ads (engineers always do this, don't they?), find ads that specifically ask for either structured or object oriented software designers.

    (a) What qualifications do you think that the potential employer really seeks?

    (b) Does the reference to a method indicate a theoretical interest in analysis, or a sizable investment in some new, automated design tool?

    (c) If you were a consultant and a customer insisted on SD (not object oriented), what specifically would you undertake to achieve this goal? What additional questions might you ask to further draw out customer requirements?

chapter four

Graphical methods

This chapter discusses graphical design methods and tools that implement some of the theory of the preceding chapters. The rules consist of the icons and logic checks for each type of graph. These include state, timing, class, and object diagrams. The discussion of automated tools presents examples of requirements engineering, simulation, documentation, and computer aided software engineering (CASE).[1] The latter is relevant because of the need for common software and system design tools. The last topic is a simple example that covers many of these aspects.

4.1 Diagram types and rules

This discussion examines rules for constructing the key diagrams presented in the last chapter. These include the STD, class and object diagrams, as well as module and process graphs. Although other diagrams are important for an actual design project, these three areas illustrate the span of activities from the top level capture of behavior, detailed functional design, and hardware allocation.

4.1.1 State transition and timing diagrams

An STD is an important starting point from which the analyst can characterize the behavior of a system. The rules for building the STD are very simple. Circles represent states and arrows depict actions that cause transitions among states.

A state is a stable configuration of the system. The analyst must have a top level decomposition to determine the elements that comprise the configurations. The class and object diagrams are methods to develop these elements so that the various system diagrams must be developed in parallel with the STD. A well-established mathematical formalism is the basis of the STD approach with its roots in both control theory and the theory of computation. Specifically,

> *A state diagram is a directed graph in which each node*
> *corresponds to a state of the machine and each directed arc*
> *indicates a possible transition from one state to another.*[2]

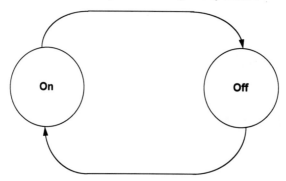

Indicated temperature > target temperature

Indicated temperature < target temperature

System component	State 1 (Heat)	State 2 (Off)
Heating element	On	Off
Fan	On	Off

Figure 4-1   STD for a heating controller.

Notice in this definition that the word "transition" is often implicit in the diagram name since without transitions the state definitions are meaningless. Although "state diagram" and STD are used interchangeably in the literature, this book will consistently revert to the more explicit STD nomenclature. Also, observe that the transitions are represented as directed graphs, which means that the sequences may not be reversible. Finally, the theory demands initial, or boundary, conditions that must be represented as part of the STD.

Consider the simple example of a thermostatic control loop. Three sensors provide inputs. The first sensor indicates ambient temperature while the second indicates the switch position. The available switch positions might be off, fan only, heat, and cool. The final input is the user selected target temperature. Consider the STD for the situation in which heating is selected, a diagram for which is shown in Figure 4-1. The start-up diagram is omitted for clarity.

The purpose of the control system in this configuration is to start the heating element and fan if the temperature falls below the selected temperature. Therefore, the two top level states are simply on and off. Transitions between these states are triggered by the indicated temperature falling below the target temperature. The electronics perform this comparison continuously.

It may be helpful to discuss the formulation of the states in this discussion. The fundamental issue is what hardware systems the control loop operates. As stated previously, in this example the loop operates the fan and heating element. These components are simultaneously either on or off, which defines the systems states as a state vector, as shown at the bottom of

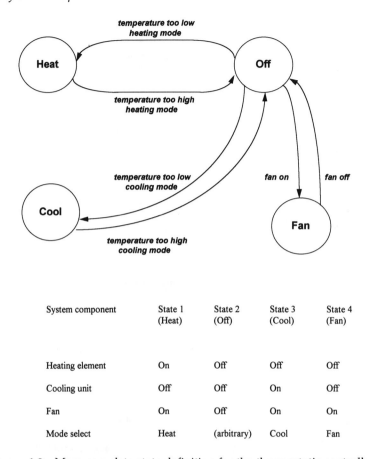

System component	State 1 (Heat)	State 2 (Off)	State 3 (Cool)	State 4 (Fan)
Heating element	On	Off	Off	Off
Cooling unit	Off	Off	On	Off
Fan	On	Off	On	On
Mode select	Heat	(arbitrary)	Cool	Fan

Figure 4-2 More complete state definition for the thermostatic controller.

Figure 4-1. If the two components could be in different states, all four combinations generate four states instead of two. Also, if the other mode inputs are allowed to produce states, the STD contains states generated by the cooling system (Figure 4-2). This figure illustrates that enumeration types, not simply digital logic, may be necessary to define states. However, in practical terms, the enumeration types activate on the basis of discretes so that two binary, digital signals could replace the four types.

The top level diagram evolves to a detailed STD by incorporating similar state diagrams for lower level components such as (assuming forced air, gas heat) the temperature comparator (thermostat), gas flow valve, electronic ignition controller, and fan controller. This also assumes that the system definition includes only the furnace and thermostat, not other elements such as the gas meter, electrical fuse box, electric meter, or other elements. These observations demonstrate the importance of clearly defining the boundaries of the system before diagramming.

More complicated diagrams require some simple rules to allow checking of the connections between states and transitions. As just suggested, the first rule is to establish the boundaries of the system. The second step involves

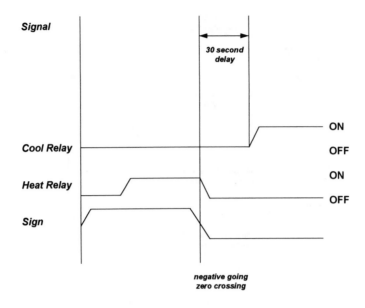

SIGN = sign of (target temperature - indicated temperature)

SIGN	Heat	Cool
- to +	trigger on	trigger off
+ to -	trigger off	trigger on

Figure 4-3 Timing diagram for avoiding race conditions.

defining elements and states. Once this step is completed, each state must be reachable in the diagram through one or more transitions.

Figure 4-2 illustrates another possible problem. Notice that all the transitions are routed through the off state. This is one method to avoid a *race condition* between heat and cool states. A race condition occurs when different states trigger on the same event but no sequencing logic is provided. Sequencing might include a time delay or a dead band around the target temperature. In this example, the event is the sign change of target temperature minus the indicated temperature. Since the hardware is unlikely to trigger perfectly off this sign change, the two units may be on simultaneously. When this occurs, the two units will compete in a very inefficient manner. The STD checks must include race conditions. One method of verifying safety bands for such conditions is a timing diagram, such as that shown in Figure 4-3. This figure illustrates the approach of using a delay, vs. the alternative concept of routing all transitions through the off state, as in Figure 4-2, which inherently should induce a delay.

Automated tools are available that can simplify the development and checking of the STD. Numerous integrated tool sets offer the capability to develop various diagrams, including the STD, as well as behavioral simulations in a standard language such as the VHDL.

An example is StateMate™ from i-Logix. This is a powerful tool that allows a designer to build the diagrams (such as STD) as well as output a VHDL simulation. The simulator verifies the correctness of diagrams internally. This product also allows the user to define document templates that can be used to produce specifications or other types of baseline descriptions.[3] There are many other examples as most VHDL vendors offer such tools.

4.1.2   Class, object, data flow diagrams

The classes and objects associated with a system design are formulated during the analysis stage. A class is represented on the diagram with a class name, list of attributes, and list of possible operations.[4] An attribute is a characteristic that defines the class. The attribute may be a numerical quantity or an enumeration type. For example, a traffic signal has three lights (class = "light") whose values can be red, yellow, or green. In a system description, the attributes are the states. Operations are the transformations among objects. In practical terms, an operation is a function or a procedure in a software system.

Links are another component of object and class diagrams. These are the relationships among classes and objects. Returning to the natural language formalism of the last chapter, the link is the verb in the sentence. As an example:

"student reads textbook"

relates the class of students (pun intended!) to the class of books by the link "reads."

The object modeling technique (OMT) of Rumbaugh[4] is a powerful graphical method for software design. This approach uses rectangles to represent classes and rounded rectangles to portray objects. A line between two object or class entities displays the links. The links can be ordered, in which case a solid dot identifies the initial class with the line terminating at the final class. Link attributes are contained within another rectangle that hangs from the link line.

In the previous example of the heating and cooling system, each of the separate units might have very similar interfaces and general behavior. That is, both the heating and cooling units accept a temperature difference (target temperature–indicated temperature) and produce an output control signal, presumably binary (e.g., a relay), that switches on the appropriate hardware. Therefore, "temperature unit" is a general category of the equipment, or a class. The objects in this case are the heating unit and cooling unit.

The class diagram formalizes the documentation methods for a class. In order to more easily understand this formalism, it may be helpful to recall the C++ language definitions, which contain the following elements:

- A class name
- A public portion that includes both data and a subprogram specification.

```
class User     {

        int
                busy ;

        public:

                int User::check_busy (void) ;
                void User::initialize (void) ;
                int User::User_destination (void) ;

        int
                id_code ;
    } ;
```

Figure 4-4   Sample C++ class definition.

Chapter 9 contains some examples, a portion of which is shown in Figure 4-4. The variable that appears before "public" is not visible globally, but everything after "public" is visible to any program that instantiates this class. There are three functions that are visible with either integer or no (void) inputs and outputs. One might also define functions that are not visible outside the class, or *class utility* routines. An example of a class utility might be the calculation of a trigonometric output from an angular input in degrees.

In the Booch notation, there are "templates" for class, class utility, and class operations.[5] This notation and formalism is somewhat more theoretical than the presentation in product literature,[3] but the OMT notation is used in this text for simpler comparison in the subsequent design examples. However, the reader must continue to focus on the important observation that VP involves system design, which contains more than software. The class utilities are the functions (either visible or local). The class operations provide return information about the class, such as its name and utilities.

The object diagram indicates the message passing and synchronization among the various objects. The diagram also represents any external I/O such as sensor data. Once again, there are templates for this information so that the presentation format is consistent.

Figure 4-5 shows the top level class structure for the heating system example. For clarity, only the attributes are shown. The class operations will be discussed shortly. Notice that the top portion of the diagram can be read in natural language:

SELECT LOGIC samples SENSORS, and SELECT
LOGIC switches RELAYS,

The sensor interface provides the indicated temperature, commanded temperature, and the mode selection. Indicated temperature is an analog measurement that is sampled at a periodic rate to produce an output. The commanded temperature is similar. The mode selection sensor might be a four-position switch by which the user selects one of the four states shown

Figure 4-5 Top level class diagram for the heating and cooling system.

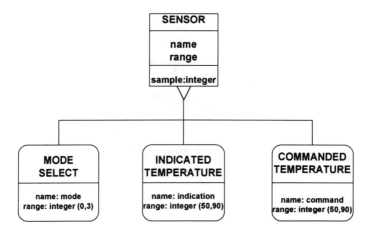

Figure 4-6 Classes and object instantiations.

in Figure 4-2. The range refers to the values that the digital outputs from these sensors can assume. The selection logic performs the temporal sequencing shown in the time line of Figure 4-3, as well as the unit selection based on the temperature difference. The relays actually turn on units as commanded, as well as verify proper operation of the selected unit.

Figure 4-6 provides more detail in the form of an object instance diagram that shows the various types of sensors. Similar diagrams are also needed for the control elements and relay systems. The reader may wish to compare this with the STD for this system that is shown in Figure 4-2. The single operation shown returns the sensor value. Once again, some details are omitted for simplicity. For example, another operation might be defined to return the sensor name, another might open communication with the sensor, another might close the channel. On the other hand, the single sampling operation might perform all of this itself using a sensor name as input (i.e., sample (<u>sensor name</u>) : integer.)

This discussion presents an overview of the formalism of object oriented design. Later examples revisit these features in more detail.

4.1.3 *DFD and process diagrams*

Another type of diagram, first encountered in the discussion of SA/SD, is the DFD. Most modern design formalisms also use the DFD despite the objections of some OOD purists that such methods are actually a composite of

Table 4-1 DFD Constructs

Name	Behavior	Symbol
process	transforms data	⬭
data flow	moves data	(producer) → (consumer)
actor	consumes data	(outside of DFD)
data store	stores data	STORAGE UNIT

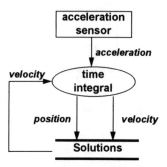

Figure 4-7   A simple DFD.

OOD and SD. Nonetheless, it is still important to map information transfers in a complex system, so the value of the DFD is undiminished for practitioners. The method consists of representing DFD elements as a named oval with named, directed arrows indicating information transfers. To avoid burdensome complexity, the transfers are identified as logical blocks of messages, not individual word transfers.

The process portion of the diagram depicts the details of data transformations in a DFD.[6] In the OMT method (described in Rumbaugh[4]), the process is shown as a named ellipse. Inputs and outputs are shown as named arrows. In addition, a control flow between processes is a selection attribute similar to an *if...then...* logic. Control flows are shown as dotted lines. The process diagram can also illustrate the relationships among hardware devices. In general, these are processing elements, storage elements (memory, disk drives), and I/O. In Booch's[5] nomenclature, the storage elements are called *devices* and the I/O elements are *channels*.

Table 4-1 summarizes the notation of Rumbaugh.[4] This table shows the name of the DFD constructs, their behavior, and the symbol that depicts the construct.

Figure 4-7 shows a simple example. In this figure, an acceleration sensor is the object. The process involves integration of an input signal with respect to time. The outputs from this process are velocity and position while the inputs are acceleration and velocity. The system stores the results of the integrations in an element called "solutions." Notice that the DFD does not indicate sequencing but only data flow. This simple case illustrates the possible ambiguity since in an implementation the two data elements marked "velocity" are actually different because they contain values at different times.

The DFD can be an important aspect of the functional allocation process that MIL-STD-499B identifies. Generally, the graphical methods provide a framework with which to accurately capture the important features of a design. This capability is an essential feature of the VP method.

4.1.4   Module diagrams

The class and object diagrams, which are, in principle, identical for all possible implementations, represent an abstract model of the system. It is equally important though to capture the relationships within a product baseline. The module diagram depicts a single hardware and software module combination. Generally, this represents a single program that runs on a hardware element, such as a data processor board. However, such diagrams can be hierarchical, as with previous diagram types, so that a module diagram can also show an entire subsystem.

Module diagrams are a mechanism to split a complex design into multiple subelements.[7] Identical classes that appear on different module diagrams are the mechanisms that link the system design. The decomposition of the system design into modules might be based on many factors, such as the hardware resources available (I/O or processor throughput, memory size), or might be based on control or data interactions. Consequently, the module diagram can show both hardware and software dependencies.

The hardware dependencies might refer to resource usage. This might involve I/O, storage, processing, or other types of devices. This perspective is similar to a diagram of a computer network broken into subdiagrams for each computer on the network. The subdiagrams correspond to module diagrams.

An example of software dependency might be a C++ "include" statement, an Ada "with," or any control information that might relate to concurrent execution. This diagram is also helpful in showing compilation order of top level software units.

In system design, the module diagram approach is the method for performing functional decomposition. It also is useful in characterizing a product baseline once candidate architectures emerge.

4.2   Vendor tool taxonomy and examples

This topic discusses examples of automated tools for requirements engineering, documentation, and simulation. These tools are available from numerous vendors hosted on popular Windows, Unix, and Digital Equipment work stations.

The underlying technology that vendors use to automate the design environment is similar to the interactive, high resolution color graphics systems found in flight simulators and video games. This allows a designer to avoid labor intensive, noncreative tasks in favor of the "game playing" that is the essence of trade studies conducted in the VP paradigm.

An important contributor to this technology is the consortium of General Electric, Marconi Advanced Concepts Center, and Interactive Development Environments (IDE). This portion of the discussion focuses on examples of such tools, including descriptions of the modeling capabilities from this consortium.

As always, the capabilities in this area change rapidly, so that each project must carefully study its options before selecting design methods. In addition, a program manager must not allow the automated tools to define the design process. Instead, the design philosophy must be the basis for selecting such tools. Despite the fact that this discussion is a snapshot, it offers a glimpse of the power of such methods and the concept of VP. This capability forms the basis of the discussion in the next chapter on the technology of VP.

4.2.1   Requirements traceability

Requirements traceability is an essential management ingredient that reduces design risk on a large program. In recognition of this, military program and engineering specialty area managers are required to maintain a database of design requirements:

> *All requirements must result from a decomposition or synthesis of the higher level requirements defined in the operational requirements. The requirements are analyzed for traceability to determine if any of the higher level requirements were not decomposed, or if any lower level requirements are not linked to higher level requirements.*[8]

As discussed previously, MIL-STD-499B classifies requirements in terms of interface definition, performance, or functionality. In addition, each requirement must be testable. Traceability refers to relating top level and low level requirements, as well as the test procedures that confirm the correct behavior of these requirements.

The numerous contributors of requirements in a large program can lead to chaos without a well-defined process. Figure 4-8 shows a "requirements tree" example for an aircraft system. The operational requirements, such as element O1 in the figure, define needs of the end users, who might be air crews, maintenance personnel, or training staff. The remainder of this diagram illustrates the requirements "flow down" process in which the engineers that represent the users refine the operational requirements into system requirements. In this example, a system is a large, relatively autonomous block of the aircraft such as the airframe, avionics, engines, environmental control system, electrical power distribution system, and others. The requirements for each of these systems are further broken down by the system integrator (with customer approval) into subsystem requirements. Avionic subsystems are often hardware implementations of functional areas such as navigation, communication equipment, displays, and flight control comput-

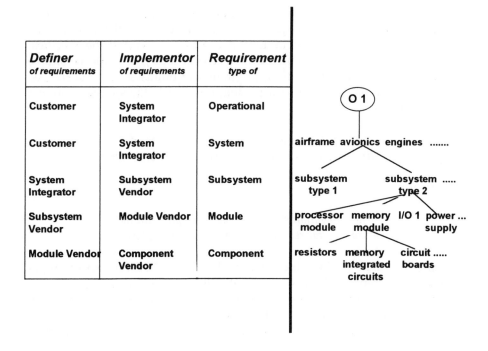

Definer of requirements	Implementor of requirements	Requirement type of
Customer	System Integrator	Operational
Customer	System Integrator	System
System Integrator	Subsystem Vendor	Subsystem
Subsystem Vendor	Module Vendor	Module
Module Vendor	Component Vendor	Component

Figure 4-8 Requirements flow down example.

ers. Computer and software designers must further decompose system requirements into functional element requirements. There may be many different types of hardware elements, as the diagram shows. The lowest level structure can be quite complex since in principle each subsystem could have different designs for hardware functional module types, but economy of scale usually forces some commonality among subsystem designs. Finally, the components from which the modules are assembled must also meet a variety of requirements.

The requirements may mature or change altogether as the system design emerges. In addition, it is important to communicate findings of specialty engineers. For example, if a design engineer working on flight control computers discovers that a commonly used capacitor has become obsolete and will not be available in two years, it is imperative that all other subsystem engineers eliminate this component from their own design as soon as possible. Finally, the various design levels must be tested to prove that they meet the requirements. Incorrectly formulated requirements may make such testing quite expensive. Regardless, the requirements flow down must be carefully documented to assure proper test coverage so that the dependencies among requirements are evident. Therefore, with changing requirements and many designers and test engineers involved, organization is important to avoid the potentially correct impression that the development is in chaos. The trail of this flow down process must be carefully documented and maintained for each operational requirement down to the lowest level requirements.

A requirements matrix is one method for tracking this process. This matrix is usually based on requirements taken from specifications at each level such as system, subsystem, module, and component. If the specifications are available in electronic format (see next topic), it would be simple enough to write a text formatting program that builds such a matrix. Assume, for example, that each requirement is stated in the specification in a manner such as:

> *The abc (topic name) shall comply with the def*
> *(requirements area) requirements of xyz (reference name).*

As an example,

> *The navigation subsystem shall comply with the*
> *environmental requirements of MIL-E-5400, Class I.*

This formulation allows the automated tool to build the matrix to simply look for occurrences of the word "shall," then parse out the subject and predicate components of the sentence. In fact, this is precisely how inexpensive requirements traceability tools work. Real-world specifications are unlikely to be so cleanly crafted, though, so a more sophisticated program is beneficial.

More sophisticated programs function as linked records in a database. When a design change forces the modification of a top level record value, the designer can sort on this record to show how this change ripples down through the levels. In the requirement example shown above, a change from MIL-E-5400 Class I to Class II would produce a dramatically large output of records modified, including other systems besides avionics. However, such a change might seem harmless to the subsystem designer and customer project lead, so the requirements matrix offers an important method to communicate the impact of design changes.

The Requirements Traceability Management™ (RTM) design tool from General Electric is an example of the mechanization of this process method. This tool provides a schema by which requirements can be "stripped" from electronic file copies of specifications in common formats such as Interleaf or Framemaker. Requirements can also be entered manually. Once a user enters a set of requirements, RTM can provide a traceability matrix similar to that shown in Figure 4-7. This analysis helps to identify those requirements branches that must "expand" (provide more details) or "focus" (eliminate needless requirements). In addition to relating requirements from different tiers, RTM also tracks the specification name, version number (or date), specification paragraph number, and author of each contributor of a requirement. RTM can also provide outputs to other automated analysis methods, such as Software through Pictures® (StP) and Cadre's Teamwork®.

Requirements traceability formulation is important in system design regardless of whether OOD or SD is the method of choice. This method allows requirements to be linked to both end user needs and test plans, so this approach offers insight into program risk as requirements change. This is an essential tenet of TQM.

4.2.2 Document and code generators

The previous discussion on requirements may suggest the scale of the task of system documentation. Each module may have dozens of components, each subsystem may have many different module types, and each system will likely involve dozens of subsystems. Therefore, a complex system may require hundreds of user's manuals, installation guides, specifications, and other documents that require periodic updates as the development progresses. Each document may be more than a hundred pages in length, so that entire system hardware documentation may consist of *tens of thousands of pages!* Of course, the software documentation may be equally voluminous. Entry of this information manually is expensive and time consuming. Once again, though, design automation can rescue the development team.

Documentation tools often require that the user develop *templates* for the output. These are based on the customer documentation format requirements. The templates can be reused from one project to another after development.

Some design tools allow the user to produce documents from graphical inputs such as the STD or other diagrams (refer to Section 4.1.1, especially the discussion of StateMate™). Others may produce outputs from behavioral simulations, such as those described in the next topic. Regardless of the method, though, a user can produce electronic files containing draft documents after a few minutes of processing. Although these documents usually require some polishing, the documentation tool can quickly provide a return on investment.

As mentioned several times, many of the design systems offer facilities to generate source code. This code might be used for simulation or as an application program. Two examples provide insight into the power of such tools. One instance is the simulation code generator offered by Synopsis. Another example is the tool set offered by General Electric as part of the OMT software.

A code generator produces simulation software from text files or graphs. Such products are now available for system level design using graphs. For example, suppose that a new data bus protocol is under investigation. It is possible to enter the STD for the bus logic and synthesize a behavioral simulation that can check detailed logic flow, timing, and sequencing. A product example is Express V-HDL™ from i-Logix, which permits STDs drawn in StateMate™ to generate VHDL code.

The OMTool™ from General Electric is a tool that allows designers to enter and verify object diagrams. C++ source code can be generated directly from these diagrams. In addition, a post processing tool can generate Structured Query Language, a standard interface for manipulating databases.

A related tool is StP from IDE. This tool provides an on-line, graphical interface for the entry and refinement of object diagrams. Once an object diagram is completed, another tool from the library can generate Ada source code.

Automatic code generators can represent a tremendous labor savings for software developers. Such tools allow designers to focus on methods and algorithms instead of code entry. Nonetheless, code generated from such tools likely requires refinement for efficiency, especially in hard, real-time systems. As always, the buyer should beware, but automated tools offer tremendous design capability.

4.2.3   Simulation

The simulation tools consist of elements such as compilers, simulators, and code generators. A compiler translates simulation code in a user file into a machine executable format. The simulator provides sequencing, control, and scheduling of simulation units. It functions as an operating system for the simulation. Finally, code generators produce compilable files either from electronic inputs of specifications or from diagrams captured with an automated tool.

Some may feel that the issue of compiler selection begs the question of simulation language. One issue is whether the simulation involves real-time execution or interactive batch processing. The latter is really the focus of the current discussion, although code generators for real-time models are certainly possible as well. Even so, C++, Pascal, Ada and other languages are popular. However, VHDL (see Section 8.3) is emerging as the preferred approach because of the powerful tools available. The language also allows the system designer to interact more closely with microcircuit designers who commonly use VHDL. This is an important concept in the VP approach, as the next two chapters describe.

VHDL simulation is an important risk reduction effort for a complex new design. The syntax of VHDL is similar to Ada or Pascal, but offers extensions to provide common simulation facilities, such as strong interface checking, as well as sequencing and timing control. Two types, behavioral and physical, are necessary. The behavioral simulation captures the logic flow, interface operation, and sequencing of the system model. A physical model adds a more detailed description of the deliverable product that might include, for example, analog electrical operation, signal pin locations and electrical parameters, and gate level operation of logic units. The behavioral model is useful in checking timing and behavior, but the physical model is needed for detailed operational impacts such as power dissipation, gate counts, size, or timing delays.

Many of the VHDL systems already provide the capability to "synthesize" physical microcircuit designs from behavioral simulations. This involves compiling logic libraries in with the behavioral simulation, as well as a more detailed description of the logic integration approach. This approach allows users to routinely synthesize fairly complex microcircuits and run application software on the physical design. The latter usually requires a hardware accelerator, though, which is often a highly efficient parallel processing simulation engine.

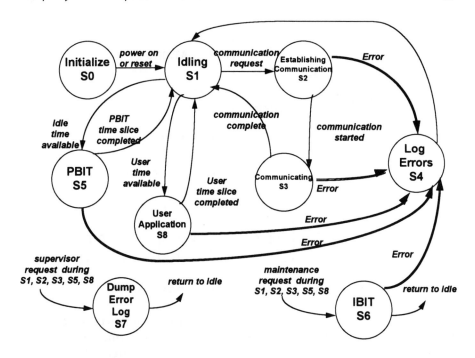

Figure 4-9   Generic system STD.

The simulator is ordinarily a software element that a simulation tool vendor can provide. An example is the VHDL system by Synopsis. This system consists of elements such as an IEEE compliant VHDL compiler, simulator, and synthesis tools. There are popular cell libraries from third parties that a logic designer can link in as part of the synthesis operation. The system level equivalent is the capability to link libraries for computer module and bus types. Manual systems configuration of this type is now possible, but can be expensive and laborious. However, an expert system could be devised to automate this process based on some standard design rules. This approach is described in more detail in Section 6.2.

Another approach is the simulation environment provided by Ascent Logic with its RDD-100® system. This system provides an automated environment for the analysis and modeling of complex systems. In addition, it can interface with many other popular tools.

4.3   Significance of the methods in VP

A brief example may demonstrate the power of both OOD and VP. The example is a top level view of an embedded computer operating system. Figure 4-9 provides an STD that illustrates the top level system operation of many embedded computers. The executive starts the various software applications on the basis of a priority pre-emption mechanism.

After initialization, the executive software begins to idle. In this example, there are both periodic built-in test (PBIT) and initiated built-in test (IBIT). PBIT is a low priority task that runs when no other software is running. PBIT

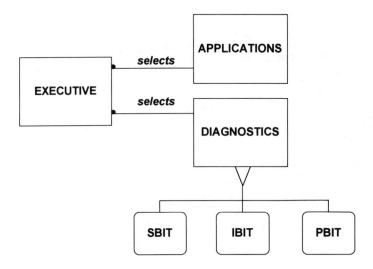

Figure 4-10   Class diagram for the previous STD.

is normally incremental since the time slices allocated for this task are small. This means, for example, that PBIT might check a small segment of memory and top level processor tests, then return at the next occurrence to continue the memory and detailed processor checks. IBIT is an off-line diagnostic routine that might be started, as an example, if a hardware error terminates the execution of the user software. The user application is another task that executes at a higher priority than PBIT. In this model, the user task starts when the system receives a message requesting a service. State S3 allows the routines in the application software intra- or inter-system communication. Communication in this model is scheduled in an on-demand fashion that might be started by the occurrence of a hardware interrupt for the I/O controller. Communication is therefore likely to be higher in priority than the user application. Finally, some type of executive must control the sequencing and is likely to be the highest priority task. An alternate approach is to poll the interface at a periodic rate to look for new information, such as changes in a status register. This approach usually requires more software overhead, though.

The STD also contains paths for error checking, shown in the figure as a double line. Errors might occur during any of the normal operations. Ordinarily, recovery from errors involves logging information so that technicians can re-create the event later, then performing a system restart. These error routines are likely to be the highest priority task, except for the executive.

Figure 4-10 shows the top level class diagram. The three classes are executive, diagnostic, and application. The interaction of these classes involves both message types for control logic and I/O for results and calculations. The figure shows the object instances of class diagnostic, but without more information about the applications, it is obviously impossible to draw the object diagram. This level of detail is inappropriate for the current discussion whose focus is process.

The next step is likely to be timing diagrams for the control flow. As suggested in the discussion of the STD, the executive, user, and PBIT code is cyclic, but communication, IBIT, and maintenance code executes on demand.

The subsequent step might be to automatically generate either a VHDL or C++ simulation from these diagrams. These simulations might help to refine resource requirements, such as execution time for each software unit, code size estimates, and I/O timing estimates. The values measured in the simulation are likely to be turned into requirements using the following comparative process:

1. Add a 50% reserve to each measured value (i.e., double the measured value).
2. Verify that the sum of all I/O and timing for the cyclically scheduled units is less than the time slice for the cyclic tasks (e.g., 50 ms for a 20 Hz schedule).
3. Verify that the processing time is less than some performance limit for the on-demand units.
4. Verify that the sum of all the code sizes is less than some limit.

These rules can then be used as inputs to an AI expert system that maps these requirements onto hardware. The expert system contains a database of hardware module types, functions performed, and capabilities. The user can feed the design requirements into the design tool, which then performs trade-offs to arrive at an optimal solution. The trade-offs are likely to be based on a prioritized list such as overall cost, reserve requirements, number of types used, and similar factors. This type of design system is the essence of computer system VP as Sections 6.2 and 6.9 describe more fully.

Many times, the priorities of these decision factors vary. For example, the cost parameter may be relatively unimportant below some limit, but grow exponentially beyond the limit. A recent method of dealing with prioritized rules of this type is called *fuzzy logic*. In fuzzy logic, each self-contained set of rules, such as the cost calculations and analysis, provides an input to some selection logic. The input is based on a *node weight* that is, in effect, the priority of the rule.

The next step might involve developing a prototype system. This is likely to include both simulated and actual hardware and software elements. For example, messages from outside the system might be generated by a simulation computer instead of actual users.

Part II provides much more detailed design examples. However, this simple example illustrates the use of automated tools as a design process. Figure 4-11 shows the three general steps in this process. The requirements process involves capturing the top level behavior and performance information. These requirements are refined iteratively using successively more detailed simulations that capture essential behavior and timing. The third step formulates and performs trade-offs for various design candidates. The final step consists of prototyping the most promising candidates. Initially, the prototypes are likely to be based on physical models synthesized from the

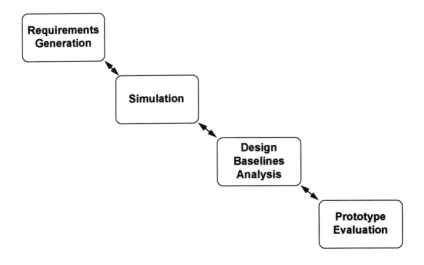

Figure 4-11   The steps of prototyping.

behavioral models. Hardware can be added incrementally to replace physical models. The next chapter explores the VP process in more detail.

References and notes

1. Kuhn, Dorothy and Sampson, Mark. *A Survey of Systems Engineering Design Automation Tools*, Proc. Third Annu. Int. Symp. Nat. Counc. Syst. Eng., July 26–28 1993, p. 279.
2. Denning, Peter et al. *Machines, Languages, and Computation* (Prentice Hall, Englewood Cliffs, NJ, 1978), p. 93.
3. Summarized from i-Logix product literature.
4. Rumbaugh, James et al. *Object Oriented Modeling and Design* (Prentice Hall, Englewood Cliffs, NJ, 1991), pp. 22–26.
5. Booch, Grady. *Objected Oriented Design, With Applications* (Benjamin/Cummings, Redwood City, CA, 1991), pp. 158–167.
6. Rumbaugh, op. cit., pp. 124–129.
7. Rumbaugh, op. cit., p. 43 and pp. 168–169.
8. *Avionics Systems Engineering Team Handbook*, Naval Air System Command, July 1993, pp. 2–7.
9. The tool descriptions of Section 4.2 are based on product information available from the marketing departments of these companies.

Projects

Goals: These projects provide examples of requirements definition and refinement.

For the following exercises, refer to the simple navigation system shown
in Figure 4-12. A linear accelerometer measures accelerations along its axis,
while an angular rate gyroscope measures the rate at which the elevation
angle changes.

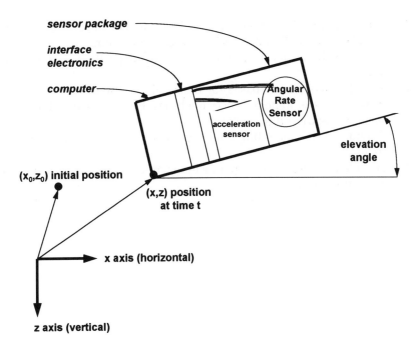

Figure 4-12 A simple position estimator.

1. Demonstrate how to derive the functional, performance, and interface requirements for the sensors and computing elements of this system based on the following top level operational requirements. Detailed calculations are unnecessary; simply show how these requirements flow down from the top level requirement. Number each requirement on the basis of the functional subelements that it affects.

   (O1) The system shall initialize within five minutes on the basis of internal elevation angle determination and external entry of the position.

   (O2) The system shall not exceed an error in position of more than one nautical mile after one hour of operation in worst case conditions.

   (O3) The sensor limits are 3 times the acceleration due to gravity at the equator for the linear accelerometer and 10 rad/sec for the rate gyroscope.

   (O4) The operator of the system shall be capable of entering position waypoints asynchronously in order to re-establish the precision of the position estimate.

2. Describe a test scenario that might be used to verify compliance of the system with each requirement of the first exercise. Number each test as in the first exercise.

3. Draw a requirements matrix ("tree diagram") that illustrates the relationships among the various requirements and test procedures.

4. Draw a top level STD of the system for operation after initialization. Include a single error state: position precision error.

5. Draw a timing diagram associated with the entry of a position waypoint update. Assume that the position update algorithm is a "hard reset" in which the initial value for the calculated position is reset to the input.
6. Identify classes and objects for this system. Discuss the behavior of these as well as some instances of objects. Example: Consider sensors as a class with multiple objects.
7. Provide a DFD, including top level processes (Hint: Refer to the STD states).
8. Provide an object model that includes modules for a possible implementation of this system.

chapter five

VP metaphors

A metaphor often captures an idea faster than repetitious, formal definitions. The mental snapshot that the metaphor creates provides the intuitive sense of the concept. For this reason, it is helpful to consider problem domains where automated design processes are more mature than for computer system design. The examples of this chapter are not VP implementations, but rather design activities that illustrate key process requirements and defining characteristics of VP.

Subsequent chapters provide more detail on the methods, tools, and opportunities of VP. This qualitative approach provides the touchstone against which the reader can gauge the significance of each of these elements. The enabling technologies include:

- The use of widely available, high performance work stations
- An intuitive, graphical interface that is easy to learn and use
- The capability to incrementally add model fidelity at a given level vs. the need to extract the most vital characteristics for the next level
- Seamless transition of design baselines among participants involved in varying levels of system abstraction.

5.1 Microcircuit design

Microcircuit design is a metaphor that appears throughout the subsequent discussions since there may be no better example of the power, speed, and flexibility of computer based, automated design tools. The technology evolution in this area demonstrates design abstraction, modularity, and the impact of high performance work stations on design.

The author offers his apologies in advance to experienced designers of microcircuits for the brute force approach to the examples. The truly gifted might return to gate level depictions at each stage to squeeze maximum performance out of an implementation. However, these examples illustrate how successive layers of abstraction can provide building blocks for automated design, although the outputs of each stage may not be optimal.

Designers build up digital logic from elementary constructs called *gates* (Figure 5-1). Some widely used gate types include AND, OR and their

R1	R2	AND	OR
0	0	0	0
0	1	0	1
1	0	0	1
1	1	1	1

Figure 5-1 Logic for three gate types.

inverted outputs (NAND, NOR.) Table 5-1 shows the Boolean equivalents for two simple gates. In the early days of design, each gate type was built up from discrete components. An AND gate, for example, might require four transistors and several resistors.[1] Abstraction is already at work since one might instinctively characterize the behavior of the gate with the logic tables instead of the electrical circuit diagram of the transistor interconnects. Nonetheless, the circuit diagrams were still useful at the early stages of this technology because if a transistor failed, one had to know how to diagnose and repair the problem. Eventually, though, gates were assembled in sealed packages that a technician could not repair. The circuit design inside the package became useful only to the electrical layout team, not the end user.

Emergence of first generation microcircuits made it possible to put many gates into a single package. An example is the adding circuit shown in Figure 5-2.[2,3] The electrical behavior of the two input bits, R1 and R2, might be controlled with various shaping circuits to comply with voltage levels, rise and fall times, and current levels of some standard, such as *transistor-transistor logic* (TTL.) The various gates add these two together in a binary representation to produce the sum at the output stage. The triangle symbol is an inverter whose function is to convert a logic 0 to a 1 and vice versa. The *carry bit* indicates overflow of the addition. A single output bit can only represent the values 0 or 1. However, if both inputs are 1, the sum is 2. The carry bit allows the designer to identify this overflow situation. Otherwise, the carry bit is 0. The block diagram at the bottom of Figure 5-2 is an abstraction that

Table 5-1   Instruction Set for the Example

Format	Meaning
add R1, R2, R3	Add the contents of registers 1 and 2, place the result in register 3
load R1, R2	Move the value at the memory location that R1 contains, place the value in register R2
store R1, R2	Move the value in R2 to the memory location that R1 contains
push R1, R2	Store the value located in scratchpad memory at pointer R1, store the result currently in R2
pop R1, R2	Fetch the value located in scratchpad memory at pointer R1, store the result in R2
branch R1	Loads the instruction pointer located in R1
return	Stops execution

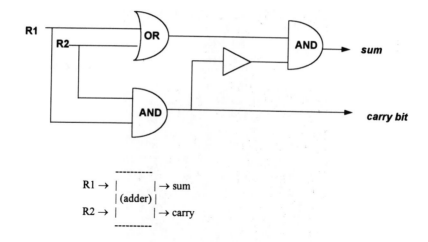

```
            ----------
    R1 → |          | → sum
         | (adder) |
    R2 → |          | → carry
            ----------
```

Figure 5-2   Single bit adder.

shows the two inputs and two outputs. Once again, when this abstraction became a physical reality in terms of package and component availability, the internal workings of the adder became unimportant to the user. The electrical designer of an adder microcircuit is still concerned with gate layouts, but is probably unconcerned with the transistor interconnects lurking one layer of abstraction lower. The design violates this general trend only when a component layout involves so many transistors that it can no longer fit within a single package. The design task defines the relevant boundaries and interfaces of each system.

Figure 5-3 illustrates a three input adder. This involves cascaded single bit adders with additional carry circuit logic. As technology miniaturization continued, it became possible to place many single bit adders into a microcircuit. This allows construction of a multiple bit adder. The logic might be built up by cascading the three input adders as Figure 5-4 shows for eight bit input registers. The last carry bit produces a new output called *integer overflow*. A designer could attach this signal to another logic unit that functions as an *interrupt controller* (more about this shortly) that can provide an indication that the sum is impossible to compute.

The designer of this microcircuit might rely on previously tested single bit adder designs, or may have some predefined, rigidly controlled interface that allows the internal design of the adder to change in parallel with the microcircuit design. This microcircuit is an example of an *integer adder*. In general, the definition of "integer" depends on the implementation. Initially, it was often 8 bits, now it is often 32 bits. The input data word length is a function once again of how much capability the packaging technology allows.

An integer arithmetic unit might include the 32 bit adder, a multiplier, support logic for subtraction and division, and an internal register set. This is a rather complex microcircuit that could be built up from logic for each of the individual operational logic circuits. Each logic unit provides a binary

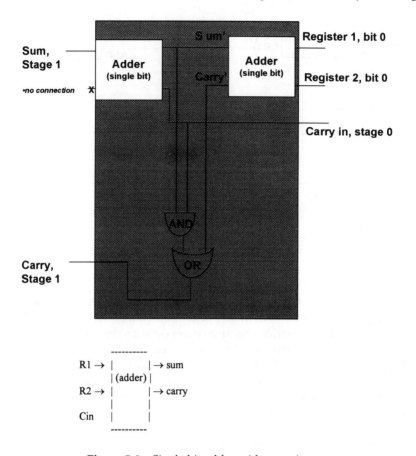

Sum,
Stage 1

Adder
(single bit)

•no connection ✗

Sum'

Carry'

Adder
(single bit)

Register 1, bit 0

Register 2, bit 0

Carry in, stage 0

AND

OR

Carry,
Stage 1

```
              ----------
    R1 →  |          | → sum
         | (adder) |
    R2 →  |          | → carry
         |          |
    Cin   |          |
              ----------
```

Figure 5-3   Single bit adder with carry input.

operation, so the circuit requires two input registers. In addition, an interface must be provided for the external control unit. This control unit must switch between data and command inputs, as well as select an *instruction* type (the integer operation) to execute. Finally, the design includes an output register and a status register. The status register contains bit patterns that indicate any error conditions. Figure 5-5 shows a block diagram of this unit. This figure also demonstrates that this type of unit leaps into the next level of abstraction since any thought of detailed transistor switching has vanished from the discussion.

A *data processor* is a general purpose computing engine. This might include several types of execution units such as integer arithmetic logic unit (ALU), memory management unit (MMU), floating point unit (FPU), and instruction decode unit. Once again, with the previous generation technology, each of these units might have been physically implemented in one or more microcircuits each. However, for modern designs, all of the units are contained within the data processor microcircuit. Notice again how the level of abstraction has risen. In fact, the behavior of each unit is sufficiently complex that the notion of *architecture* emerges. The architecture is the rela-

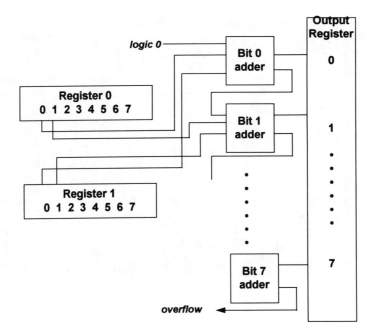

Figure 5-4   Multiple bit adder.

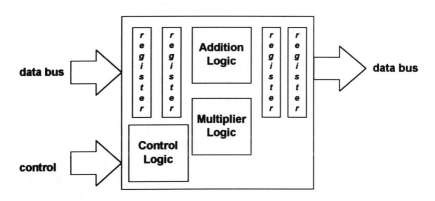

Figure 5-5   Integer arithmetic unit.

tionship among the various constituents (logic units) that is the basis of the system level behavior.

The instruction decode unit may deserve additional discussion. The communication interface to the memory system is a *memory bus* that includes a data path, an address path, and additional control signals. One of these control signals indicates whether the bit pattern contained on the address channel points to a *data segment* or a *code segment*. The data segment simply contains 32 bit words that represent values. However, the code segments of memory contain bit patterns for instructions. If an input is data, the instruction execution unit simply loads the input registers, but if the input is an

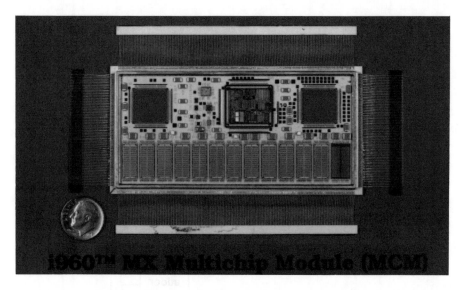

Figure 5-6   ATK-32™ MCM example. (Photograph courtesy of Alliant Techsystems.)

instruction, the unit must decode it and schedule the hardware resources necessary to perform the instruction. As an example, consider an integer addition that one might symbolically represent as:

add R1, R2, R3

This means to add the contents of general purpose registers 1 and 2 and store the result in register 3. Of course, this symbolism means nothing to the hardware, which recognizes only the bit patterns for the instructions. Section 5.2 explores the relationships among the hardware and software layers, so it is redundant to detail this symbolism any further now. However, these observations demonstrate how abstract the representations have become in comparison to the original descriptions of transistor operations.

Recent hardware technology advances make possible even higher levels of integration. The multiple circuit module (MCM) is a very large microcircuit, typically more than 2 in. per side in length and width. Figure 5-6 shows an example in the form of the Alliant Techsystems ATK-32™. This MCM contains an Intel 80960MX™ data processor, random access memory, electronically erasable programmable read only memory, and several communication interfaces.

As the microcircuit technologies became more complicated, designers used abstract models to manage the complexity. As an automated design method, one can capture the detailed timing and functionality of these abstractions as a *behavioral simulation*. These simulations can run in parallel in order that designers are able to wring out any anomalous interactions. Once this is complete, the designers can fabricate instructions for manufacturing the product. The automated process of performing this step is called *synthesis*. The resulting electronic files are called *physical models*, which contain hardware definitions such as signal pin assignments and locations on the microcircuit. The last step of the automated manufacturing process is generation of the *Netlist*, which contains software instructions for numerically controlled machines that produce microcircuits.

Several characteristics key to microcircuit design are also essential to VP. The first aspect is high performance computer technology at a modest price. Current implementations that host such systems might be high performance work stations networked into a cluster, but future systems may involve a spectrum of massively parallel performance engines with networked work stations and stand alone systems. Another aspect is the use of an intuitive, graphical interface. This replaces older design drawing methods that involved laborious, manual design entry and checking. Finally, this technology also permits reuse of both requirements and design elements. Reuse is also a key factor that can make VP cost effective.

This process allows microcircuit designers to work with abstract models that limit the complexity of the design problem. The essential features of each model can execute concurrently with other relevant models to verify the correct operation of the unit under testing in the operational environment. The automated tools allow the designer to pull in physical implementations of various unit types from software libraries without the need to worry about the detailed electrical implementations that are reusable and usually change very slowly over time.

If one thinks of this process in the most general terms, this description is precisely the requisite design environment for VP. This allows only the most important features of a computer design to be modeled initially, with more detail added as the design matures.

5.2   *Software design*

Software design offers another model for VP. The software process models and formal design standards couple with similar systems engineering standards to form an integrated hardware and software development team. Subsequent discussions point to the need for common process models and formal methods for both software and system designers in order to avoid a replaying of the current generation "software crisis" in the next generation of massively parallel and highly integrated systems.

Software methods grew in abstraction over time to deal with the need for increased complexity. This evolution mirrors that of computer hardware that Section 5.1 describes. For the sake of clarity, this section avoids the chronological references of the last discussion and instead describes the levels of abstraction in terms of software structure. This begins at the hardware level and extends up to the software application.

At the hardware level, the instruction stream is literally ones and zeroes. The memory locations that store these instructions and the signal lines that connect them to the execution engine have only two stable states. Although "on" and "off" might be more appropriate for a hardware discussion, the conventional nomenclature for these states is "one" and "zero." The sequence begins each time that the execution engine increments its instruction pointer, which indicates which instruction to fetch from memory. Overlook for now how one loads instructions into memory, as this topic arises again shortly, and consider a simple example.

Suppose that the *instruction set architecture* (ISA) contains several primitive instructions that operate as shown in Table 5-1. The convention for the load and store operations is that the memory address is in the first register while the value is in the second register. Notice that the level of abstraction has already increased since one identifies instructions in terms of a name, not a bit pattern. However, this representation still requires that the user schedule hardware resources, in this case, registers. This language is called a *register transfer language* (RTL) and, for a data processor, the assembly language.

The add instruction should be familiar from the previous discussions. Load and store instructions move data from the application program storage area. Push and pop move data from scratchpad memory that the processor uses to store temporary results. The branch and return instructions control execution. The unconditional branch is usually called a *jump* in most ISAs. An actual ISA might also contain conditional branches based on the occurrences of conditions. For example, a branch instruction such as BEQ, R1, R2 symbol might represent a decision point to branch to pointer R1 if the value contained in R2 is zero. If the value is not zero, no branch occurs. The return instruction cycles the control flow back to a previously stored point of execution. If no pointer is provided, the return statement simply halts the execution of the application program.

This interface is simpler to use for a software designer than a bit pattern representation. The bit patterns represent the swings of various signal lines, but a software designer should not be forced to deal with the complexity of the detailed hardware operation. However, the assembly language formulation still contains hardware resource scheduling to a frustrating extent. Further abstraction is needed.

A *higher order language* (HOL) contains a richer, more natural, abstract symbol set. A HOL allows the software developer to define and use logical names for data elements instead of referring to the memory or register that contains the variable. Most HOLs contain integer arithmetic operators and other mathematical operations. A special software tool called a *compiler* translates the symbols in a text file into assembly language. Another tool, called a *linker*, converts the assembly language into bit patterns. In practical terms, most compiler vendors use proprietary *intermediate pass languages* (IPL), not assembly language, to produce the bit patterns most efficiently. However, nearly all compilation systems provide an assembly language listing as an output.

A *source level debugger* is a tool that allows a software developer to slowly step through the execution of a program. This tool is not a compiler, but many vendors provide on-line assembly and "dump" listings. Figure 5-7 shows an example. The memory dump shows values in base 16 (hexadecimal), which is normally more convenient to use than binary. The hexadeci-

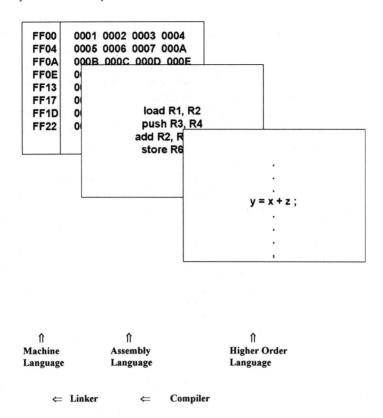

FF00	0001 0002 0003 0004
FF04	0005 0006 0007 000A
FF0A	000B 000C 000D 000E
FF0E	0
FF13	0
FF17	0
FF1D	0
FF22	0

load R1, R2
push R3, R4
add R2, R
store R6

y = x + z ;

⇑   ⇑   ⇑
Machine   **Assembly**   **Higher Order**
Language   **Language**   **Language**

⇐ **Linker**   ⇐   **Compiler**

Figure 5-7 Outputs from a symbolic debugger.

mal values in the far left column are the starting memory address for the first entry. The second window shows an assembly language listing, while the third window shows the source level HOL listing.

A final software tool is the *kernel and monitor code* that loads and starts application programs. The monitor usually is resident in the firmware of the target hardware circuit card assembly, but a communications package on the *console* moves the code from storage (e.g., hard disk drive) to the target module. Any messages from the console or application also appear on the console display, which is typically a high performance work station.

Two other concepts may further illustrate the concept of abstract models as a software interface. The first example is *virtual memory*, the second is the *virtual machine*.

Virtual memory is related to the notion of logical addressing in which software can reference a name instead of a physical address. This abstraction is equivalent to the relationship between algebra and arithmetic in which binary operators modify symbols instead of numbers. The power of this approach is that it provides a common behavior and interface for all mass storage devices such as nonvolatile memory, fixed disk drives, tape drives, or other media. Many operating systems use virtual memory for *on-demand virtual paging* in which a driver program, called a page swapper, loads the next process that the execution unit must service and stores the current process in whatever mass storage is available. A related application of a virtual memory interface is a compiler that must generate addresses for variables in the input source code. The compiler converts the source code into an intermediate pass language with predefined, general data structures called *glue vectors* similar to the programmable logic that some hardware designers call general logic units (GLU). The linker program converts the logical addresses into physical addresses.

The virtual machine is a generalization of virtual memory that abstracts all of the key features of a data processor. These could include interrupts, memory control, register manipulation, and other low level hardware operations common to most assembly languages. The performance penalty of this generalization makes it normally unsuitable for operating system usage. However, these models are valuable in many areas of theory, such as automata and computer language theory. In addition, some designers implement the model in products as a middle layer interface for certain design tools such as instruction set simulators.

This presentation shows how abstraction simplifies software design. At the HOL level, just as with microcircuits or MCMs, architecture becomes an issue. Formal methods allow designers to establish the relationship of various software modules and control behavior. Many of the methods involve automated tools to speed this process and reduce the complexity for designers. Once again, the capability for abstraction and high performance automated design tools are precisely the elements that enable VP.

5.3   Hardware hotbench

This discussion does not present another metaphor for VP, but the converse, a summary of prototyping without automation. This involves incremental fabrication of both hardware and software prototypes.

The hardware hotbench is the functional test execution capability for an actual system. Ordinarily, one conducts the environmental qualification tests separately from the functional tests. Functional tests verify that the system can meet the interface and performance demands of a customer. The environmental qualification demonstrates that the hardware operates safely in the vibration, temperature, altitude, humidity, and other envelopes associated with the system installation.

The functional test must provide all the communication features of an installation. Consider an airborne, inertial navigation subsystem. The computer produces the navigation solutions on the basis of an initial solution that the user provides at start up and from sensor inputs during the update phase. The sensors detect linear translations and rotational motion. The functional testing of such systems often proceeds in a stepwise fashion.

Initially, the question may simply be whether the software produces solutions of sufficient accuracy. To answer this question, the engineering team might provide a prototype software system with the sensor inputs "stubbed out" to accept a predefined set of values. These values might be read from a data file at each step of the calculations. The next step is to move the prototype code to the target data processor module. Initially, the sensor inputs are probably still stubbed out so that this represents a *single board test* involving only the processor module. After completing this step, the team might move to a multiple board test in which the processor module communicates with I/O modules. Initially, the I/O modules might be loaded with the scenario files for sensor data and mode command inputs. Incrementally, one must add in the hardware I/O one interface module at a time. For example, there is usually a different I/O interface for the sensor data than for the command inputs. One often replaces the track files for the sensor data with test equipment that can produce the electrical signal patterns that correspond to the data in the test file. At some point, it becomes necessary to apply actual forces to the navigation subsystem in order to stimulate the sensors. Once again, there may be several steps in this testing. The team might install the subsystem in a ground based vehicle and drive it around for many hours to test the accuracy of the results. This verifies long-term stability of the solutions but is likely to use only a small portion of the full sensor range. The developers might use a motion table to explore the sensor range. Finally, the subsystem might be installed in an aircraft for additional testing.

This scenario is typical for many types of embedded equipment. Another example might be an airborne navigation radio that the crew uses for area navigation. The laboratory testing might involve test equipment as the last example discusses. The developers might follow this with ground based tests by placing such radios in small vans and comparing the indications to ground based reference points. The functional tests usually always end with some type of flight testing prior to the start of subsystem production.

Consider this testing from the perspective of an experienced program manager whose concern is a quick recovery if problems emerge. Where might unexpected difficulties arise? First, the team adopts the test process described above so that the steps are small enough to present acceptable risk. Yet, each step is a leap of faith. In moving from the prototype code analysis to execution of the single board test, the behavior of the target hardware module may be sufficiently unexpected and different from the analysis that the code simply does not execute properly. This is often a problem with new designs. Subsequent steps may also provoke excitement. The issue is whether a method exists to smooth the steps of this test process any further.

One source of the difficulty lies in the need to produce a hardware prototype before any software testing can begin. If the hardware proves to be inadequate, the software designers must return to requirements analysis while the remainder of the team sits idle. Imagine instead that the design files that the hardware team uses as a product baseline are also available to software developers. The systems engineers can assemble these files, with the assistance of automated tools, into a VP. The software team can execute prototype code on this VP early enough in the design process to still have an impact on hardware requirements. This is the essence of avoiding a repeti-

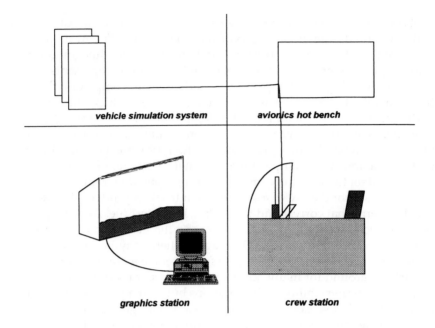

Figure 5-8   Elements of a flight simulator.

tion of the software crisis. In addition, it can help to avoid costly fabrication cycles of early prototypes.

5.4   Flight simulation and the avionics hotbench

Flight simulation can provide a high fidelity environment for testing the navigation computer of the last topic. Instead of track files, the aerodynamics of the simulated aircraft can produce sensor values. This approach is called the *virtual environment*. Therefore, it offers an interesting example of the interrelationships among the virtual and actual approaches.

A spectrum of uses for flight simulation systems exists, from aircrew trainers, to equipment test stations. The elements of flight simulation used in embedded computer design and testing (see Figure 5-8) are the graphics system, cockpit work station, vehicle simulation host, and avionics hotbench.

The graphics system provides the "out the window scene," which is a depiction of the ground, sky, and features. For many laboratories, the graphics software runs one or more high performance work stations. This software might load in a database that contains (x,y,z) triplets of latitude, longitude, and terrain elevation. The separation among the data points is usually 100 m (330 ft) and the more sophisticated systems can smooth the data over intermediate areas. A 19-in. color display serves as the primary output but additional optical devices can expand this into a wide screen image, usually on a 14-ft screen. Multiple work stations may be necessary if side views are also required. These systems can produce impressive images capable of night scenes replete with twinkling stars and automobile headlights on highways, daytime images that include the sun, clouds, shadows of objects on the ground, and many other characteristics. Although these systems do not produce the photographic quality of scenes used for pilot training, they are highly programmable at a cost typically 10% of a high-end system.

The cockpit work station provides the controls and displays interfaces for the operator. Typical controls are throttles, rudder pedals, brakes, and a control stick. Often these are instrumented with a linear potentiometer whose position is read through an analog to digital interface for the simulation computer. For some applications, especially pilot training, there may also be a *force-feel system* installed. This provides programmable control forces so that the software engineers can simulate the changes that an aircraft presents because of differences in airspeed. For avionics testing, this system might not be essential, especially in consideration of the high cost of such devices. The displays interface consists of various programmable switches, common analog instruments (such as altimeters or vertical speed indicators), and graphics displays. The latter might be a cathode ray tube or some type of flat panel display. For a general purpose laboratory, the displays might be of a general purpose design, but a laboratory dedicated to a single platform might use the actual aircraft displays.

The vehicle simulation host is normally a mini-computer or main frame because of the performance necessary for the vehicle dynamics and environment. The vehicle dynamics include the six degree of freedom calculations as well as control system models that couple the operator inputs to the dynamics. The environment includes the atmosphere model for air density, temperature, pressure, and other variables varying with altitude. Another model, often called "ground," might modify the aircraft dynamics when the difference between the terrain elevation and the aircraft altitude becomes zero. This model is necessary for take-offs, landing, and simulation of terrain impact. Other models might be present for threats and their electromagnetic radiation patterns.

Finally, the avionics hotbench is similar to the test benches that the last topic discusses. The hotbench provides mounting hardware for avionics that is similar to aircraft systems. It also provides electrical power and wiring harnesses for digital communication channels. A primary role, though, of the hotbench is to provide programmable test equipment that can stimulate the sensor interfaces of the avionics. For example, the unit under testing might be a radar warning receiver, so the simulation team could connect a wide band pattern generator behind the sensor heads to replace the antennae and front end electronics. As the operator flies the simulated aircraft through a scenario, threats "pop up" in the software and trigger the test equipment. This capability provides a highly repeatable set of test conditions that are much less expensive than flight tests.

Virtual reality technology sparks new interest in the *many on many encounter*, which is a classical simulation problem. Originally, this involved air combat, such as several simulated fighter aircraft engaged in battle. More recently, this might involve integrated air and land forces at many different simulation laboratories. Imagine that the simulation begins with an armored thrust against a well-entrenched force. The defensive forces call in counter air strikes to halt the advance. As the simulated attack aircraft begin to destroy the armor, the enemy might call in its simulated counter air fighters to sweep the skies. These simulations can offer practical lessons to battlefield commanders and lowest organizational unit leaders about the response times and capabilities of various types of forces available to them. This *virtual battlefield* capability can save both dollars and lives in training exercises.

5.5 Summary

The previous chapter explored the relationship among software and systems engineering process models and management requirements. The next chapter, with its emphasis on OOD, illustrates the need for common software and system design methods. The metaphors of this chapter offer some images of the challenges to be met in integrating complex systems.

The metaphors also illustrate how technology meets methods to provide capability. The cost effective, high performance work stations open the door of virtual reality to many applications that might have demanded supercomputers a decade earlier. These work stations present an intuitive, graphical interface that does not demand specialized knowledge of the host system operation. The performance of these systems also allows multiple players, even huge, wargaming simulations running in real time.

Another aspect that these metaphors demonstrate is the incremental addition of fidelity and detail in simulation models. A design team can start from the most essential features and then add more complexity as a more complete test environment is desired.

An abstraction of this process is easy to capture in a simple model. During the design effort, the iterative approach to developing both the requirements and the product baseline is essential to capture in a model. Another feature is the capability that work stations and modern software offer at a modest cost. A final feature is the interaction of hardware and software process models and formal methods. These elements are shown in the process model of Figure 5-9. This model is the basis for many subsequent discussions.

Periodic market surveys allow technologists to establish a baseline description of the capabilities and suppliers for key components. In a computer system, these include the obvious elements such as data processors, memory, and I/O controllers. Equally critical are support devices such as timers, interrupts controllers, and application specific integrated circuits (ASIC). The surveys must characterize the performance of such elements, but many other factors are also important. These might include trade-offs between product maturity and obsolescence or complexity vs. fault tolerance and data security. This information is the basis on which a systems engineer can establish achievable technical requirements that are compatible with customer needs and achievable technology.

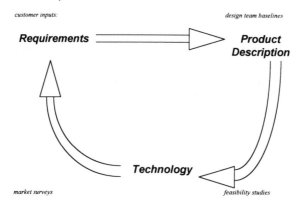

Figure 5-9   The cycle of VP.

The detailed implementations of the requirements emerge as product baselines. The design team must evaluate the candidates against the technology criteria just presented, cost, and schedule. Formal feasibility studies are often the basis of such selections.

Process models are discussed in Chapter 2. One should recall the distinction between spiral and linear models. Figure 5-9 represents a spiral model that is consistent with historical results from many large programs. Perhaps the significance of VP might be best captured in the observation that one might model each event between these three pillars as an electronic file transfer. The technologists might e-mail the results of a market survey to a requirements team. The design team might e-mail product baselines to the technologists (or a validation group) for evaluation. A program manager might also require that customer requirements and design baselines be represented in some standard file format. The validation group might use these files to configure simulations or test scenarios for prototype hardware. This is the essence of VP.

References and notes

1. Mukherjee, Amar. *Introduction to nMOS and CMOS VLSI Systems Design* (Prentice-Hall, Englewood Cliffs, NJ, 1986), p. 65.
2. Fletcher, William. *An Engineering Approach to Digital Design* (Prentice-Hall, Englewood Cliffs, NJ, 1980), pp. 200–203.
3. Diefenderfer, James. *Principles of Electronic Instrumentation* (W.B. Saunders, Philadelphia, PA, 1972), pp. 414–415.

Projects

1. Consider the microcircuit design metaphor.
   (a) Assemble some product literature for the automated design of ASIC.
   (b) Identify at least two classes of logic types. What are the defining characteristics of the objects in each class? Hint: Use the product literature.

    (c) Does the ASIC library promote reuse among various design implementations? What features of the tool set simplify reuse of primitive logic types?

    (d) Compare and contrast the design cycle for microcircuits to the diagram shown in Figure 5-9.

2. Consider the software design process.

    (a) Compare and contrast the iterative model of design shown in Figure 5-9 with the CMM of Chapter 2.2.

    (b) Discuss the applicability of an ISA simulator. Some vendors can provide the complete symbolic debugger capability with the simulator, but there are limits to the hardware support features included. Provide a summary of an actual product of this type.

    (c) Discuss some other elements that one must add to the ISA simulation to provide an emulation of a data processor module. Hint: How might one deal with interrupts and I/O?

3. Discuss how the results from a hardware hotbench can be used to validate simulations and VP.

    (a) Test coverage is a measure of the total number of inputs, outputs, and control flows vs. those that the test scenario actually exercises. Engineers often express this coverage as a percentage. How does one know whether the test data provide sufficient coverage? (Hint: Think about the probability of each control flow and its impact on the I/O.)

    (b) How can one develop input test data for a simulation? Discuss the advantages of random pattern inputs versus predefined inputs.

    (c) How can one establish the correct outputs for the corresponding inputs? How can one establish output values for the random input patterns? (Hint: Suppose that the hardware produces the same outputs as the simulation for the fixed inputs. A test engineer can attach a random pattern generator to the hardware inputs. How can one capture the outputs from the hardware? Is it valid to use the hardware outputs as the test vectors? Why?)

    (d) Develop a top level, abstract model for the hardware test bench by listing the procedures involved in running a test. How might one simulate these steps? How might such simulations be integrated into a virtual hotbench?

4. Consider the flight simulation environment.

    (a) Provide a top level analysis of the flight simulator using the control/input/process/output (CIPO) model. Identify the inputs and outputs for a computer unit under testing.

    (b) Discuss some of the behavior that characterizes the actual unit under testing. Together with the I/O of the previous question and these behaviors, assemble a description of a VP for the computer system.

    (c) Is the operator in the loop necessary for the simulation of the VP? Discuss advantages of "canned" operator control input scenarios vs. operator in the loop tests of the VP. Hint: Revisit question 3(a).

5. MIL-STD-499B breaks requirements into three areas: functional, performance, and interface. Consider validation of the requirements for a new computer design.
   (a) Which of the three categories might be the easiest to test with a first generation VP? Why?
   (b) Which of the three categories might be the most difficult to test with a first generation VP? Why?
   (c) How can one augment the test data from the VP to gain more confidence in the results?

chapter six

Technology and overview of VP

The essence of the virtual environment is a simulation that creates the illusion of user interaction with objects using natural senses, especially sight and sound. Three elements are key criteria that distinguish such systems from other design tools using high resolution graphics. Much of the theory and technology derives from AI research. The features are:

- Movement (the user must be able to move naturally in the environment)
- Communication using human senses via stereoscopic vision or sound I/O systems
- Behavior that is complex and representative of the objects involved.

For most applications, an additional feature is essential. This is the interaction between the human operator and the simulation system. The operator provides an input and the simulated system must respond very nearly in the same amount of time that the actual system would take for such inputs.

This chapter discusses the link between AI and VP. The discussion consists of two main elements. The first element provides an overview of AI technology elements. The second describes a conceptual VP design process and relates this to AI and other systems engineering methods. The description of the conceptual VP process continues in the introduction of Part II after some discussion of additional theory.

6.1 Technology elements

The technological basis of VP consists of decision making software algorithms and sophisticated sensor processing. These include synthetic vision, natural language, AI expert systems, fuzzy logic, and virtual reality. The VP software and hardware systems rely upon these elements to immerse the designer in a virtual environment that permits interaction with a proposed product baseline.

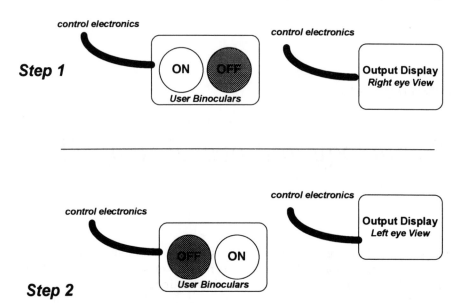

Figure 6-1   Virtual reality displays.

6.1.1   *Virtual reality*

Virtual reality refers to the generation of high fidelity, stereoscopic graphics. The three dimensional view can be generated using special display equipment, such as binocular output or special polarizing, filtered outputs for each eye. The binocular outputs can flash periodically as shown in Figure 6-1. This gives a simulation participant the illusion of interacting with the simulated environment.

One distinction between virtual reality and conventional high fidelity displays is the capability for multiple views to simulate three dimensional objects. Consider the flight simulator example. If the pilot cannot see something outside clearly in actual flight, he might move his head to obtain a better perspective. On a conventional flight simulator, this movement does not change the view. In a virtual reality system, though, the pilot might wear a sensor that allows the computer to locate the position of the pilot's eyes. This sensor might provide an input to the simulator that changes the view. For example, if the pilot moves forward to look as far back as possible out of the left window, the computer can slide this view forward as the position of the pilot's eyes change. The other pilot in the right seat might have similar sensors. If both pilots wear the shuttered binoculars just described, separate views for the captain and first officer are possible.

A more common example is the architectural design walk-through in which customers can inspect a building design using virtual reality. This allows the customer to examine window and door locations, placement of other architectural fixtures such as a kitchen peninsula, and finish features such as exterior shutter type and placement, landscaping, and similar fea-

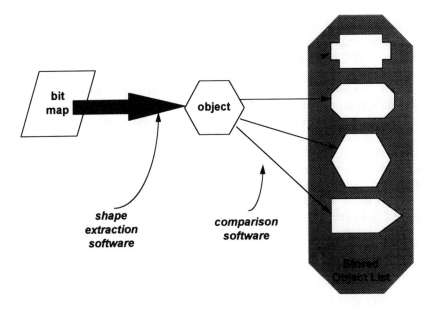

Figure 6-2 Object classification system.

tures. This approach proves competitive in cost because blueprint "redlining" can be very expensive for the customer.

A variety of other applications continues to emerge. Complex relationships among data can be plotted and viewed three dimensionally.[1] This is a possible relationship with VP for system design, as Section 6.4 discusses. Perhaps it is unsurprising that high end video games have embraced this technology as well.[2] There are a host of other demonstration and development systems that run on high performance personal computers or work stations.[3]

6.1.2 Synthetic vision

Synthetic vision is digital processing to achieve recognition of optical images. For example, if a video camera provides the input, the video signal can be periodically digitized with "frame grabber" hardware. Special purpose image processing software can then manipulate this image. An example of this might be an object classification system that compares the bit map to stored objects in order to classify the shapes. An overview of this process is shown in Figure 6-2.

The update rate for manipulation of video images will be quite low unless the software contains some sophisticated mathematics. The problem stems from the large number of data points needed to describe an object. In addition, a classification method may need several different perspectives of a three dimensional object. The software might then rotate the three dimensional stored image in comparison to sensed shapes. However, this process can be quite time consuming, even on a high performance processor. The key to efficiency is to store the minimum number of points needed to classify an object. One recent method for dealing with this problem is a branch of geometry called *fractals*.

The key technology associated with fractals is the iterated function system (IFS). IFS allows complex objects to be described in terms of iterative mathematical equations that, when executed, build up a presentation of the object. The advantage of the IFS approach is that the number of parameters involved can be much lower than the data points necessary to describe even a simple object. One classic example is the Sierpinski triangle, a sequence of dark and light colored nested triangles. The IFS method requires only 5 parameters to describe this object, but there are 40 triangles. A description of this object using the (x,y) position of each end point of the triangle would require 240 values![4]

Synthetic vision is an important aspect of AI, and therefore of VP. This technology provides a powerful sensor technology for applications such as robotics, or other hardware control systems.

6.1.3 Natural language

Natural language (NL) refers to next generation software languages that mimic spoken language instead of more arcane software languages. This capability can free the design process from details of a specialized computer language. This can speed code entry and make powerful processing systems more accessible to experts with otherwise minimal programming skills.

NL is not simply an important AI element in its own right, but also is an important element of other AI technologies. Perhaps the most important application is as a front end for an expert system. This feature allows a user to interact with the AI technology in a manner similar to communication with human experts.

Voice recognition (VR) systems have become a leading edge approach to natural language. This method captures the essence of the natural language goal, which is to allow a user to interact with the computer system and AI software as if it were a human expert. VR implementations usually are based on comparison of voice inputs to stored "key words" that can be characterized either in terms of conventional signal processing methods or comparison to phoneme tables. The latter approach is similar to the phonics method of reading unfamiliar words that some elementary schools teach. Unfortunately, each new user of a system must train it to recognize his or her voice since speech characteristics differ so widely. Many VR systems support several hundred key words. To train such a system usually requires that the new user speak each key word at least three times. With a smaller number of key words, though, recognition becomes less of a problem so that at some point, training the system might not be required. This is a design trade-off the system engineers must address.

Real-time hardware control is one application of VR that can reduce operator workload significantly. Aircraft systems control might be the first emergence of VR. The initial systems are likely to permit aircrews to change menu pages or select items for graphics displays in highly automated cockpits.

VR is not the only method of natural language. An interactive system can allow a user to learn a new subject from a computer terminal. The user enters a question in the native language and the NL system decomposes the sentence on the basis of syntax and key words. Assume that a system accepts only questions as inputs; then the user question "what time is it?" might be parsed and processed as:

```
subject = time ;
verb = is ;
......
if (subject = time) && (verb = is)
{
get_time() ;
}
```

A query for the date ("what day is it?") might evolve as:

```
subject = day ;
verb = is ;
......
if (subject = day) && (verb = is)
{
get_date() ;
}
```

The NL method refers to the processing of an input based on rules of the syntax of a spoken language. The methods vary among systems as to how the user enters the query.

6.1.4   *Knowledge based systems*

The topic of AI covers a broad scope of applications and methods. However, their essence is to simulate the problem solving techniques of a human expert to reach a solution. Although search algorithms are critical elements, the knowledge based systems differ from conventional software algorithms. The former provide a general framework or structure to problem solving to which one can add data. The conventional algorithms often do not separate the structure of the data from equations and control sequences that form the algorithm. Finally, many of the knowledge based methods rely on heuristics or phenomenology that are essentially rules of thumb or accepted practices.

An expert system is an abstract model of human problem solving. This model consists of elements such as those shown in Figure 6-3 that offer efficient implementations on computer systems. The input to the system might be an NL processor that the last topic describes. The sensors could be

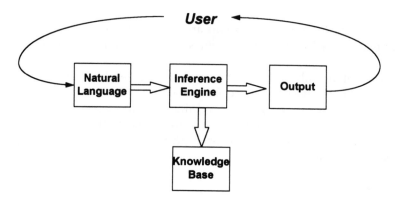

Figure 6-3   Expert system elements.

a keyboard and screen, a VR system, bar code scanners or other visual sensors, or other application specific methods. The knowledge base is the software description of problem situations and the recommendations of an expert in the field. The inference engine accepts the processed user input, then executes software logic to determine which rules in the knowledge base apply. This block then feeds the recommendations and a summary of the applicable rules to the output section. The outputs consist of the recommendations. In addition, the user might request a justification of the answer. In response, the system displays the rules and logic used to make the recommendation.

Part II explores the relationship between expert systems and VP. Several definitions and concepts lay the basis for the later discussions. The first concept is that of the *rule*. A rule is a simple logical relationship usually expressed in an IF-THEN manner. *Facts* are the input data that are logically true. In a very complex system, the rules can mesh together in subtle ways. If a user changes data, it could invalidate previously valid facts or conclusions. An expert system that requires the user to recheck the output of all rules whenever the data change is called *monotonic*. A nonmonotonic is very useful and powerful, but it may produce results much more slowly because of the more sophisticated rule checking. The *inference network* is the relationships among data and rules for an implementation and one can represent this architecture graphically. The *inference engine* is the implementation.[5]

Rule based systems can be efficient for some applications, but there are also several disadvantages to this approach. The first problem is infinite chaining. One might add a rule that effectively locks the system by placing it in an infinite loop. This often involves antecedent relationships such as:

IF (i is an integer)
THEN (i+1 is an integer)

The addition of this rule might force the rule based system to update the list of facts for each integer data entry. Another problem is the addition of a

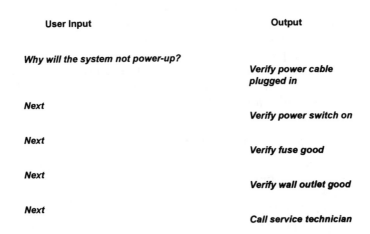

Figure 6-4 User diagnostics expert system.

new rule (or modification of an existing one) that might contradict a previous rule. However, rule based systems offer a relatively simple and intuitive method for implementing problem solving. For highly structured, modular problem domains, such systems can work well.

User's manuals and diagnostic tests are popular applications of expert systems. As a simple example, suppose that a user cannot get a computer to begin execution. The session might transpire as shown in Figure 6-4. This system might contain a variety of situations, from software diagnostic codes displayed on the screen to inability to start the system, as shown in the figure. The input and output probably involve an interactive computer system. Once the input section translates the problem, the inference engine probably just dumps recommendations from the knowledge base to the output. However, in more complicated examples, the inference engine must also contain pattern matching logic that can select the appropriate rules from the knowledge base.

The data that drive the rules can be structured as objects. This approach can bring all of the advantages of OOD to knowledge based systems. Part II explores this approach and its impact on VP.

Diagnostics are popular applications of expert systems. One of the first demonstration systems that proved successful in providing diagnosis of infectious diseases was a product called MYCIN.[6] Aircraft diagnostics are another emerging application. The financial trade-off for such systems is the cost of initial and recurrent training of technicians, time to achieve a diagnosis, and documentation vs. the expert system development and maintenance. This tends to shift the focus to complex systems that require expensive, time consuming testing. There are many attractive tasks of this nature in hardware control software, but these applications are more problematic because of test issues. Expert systems often involve large numbers of logic branches that can be exceptionally complicated to test thoroughly. In addition, some expert systems allow the inference engine to add rules based on user inputs. Although these "learning" systems are very powerful, they once again raise troubling questions about software testing and maintenance.

Expert systems play an important role in computer system VP. Later discussions describe the highly automated design process that improves design cost and time. An expert system can be an integral part of such design, especially in terms of analyzing the results of trade studies.

6.1.5   Fuzzy logic and neural networks

The previous topic discussed knowledge based systems, especially those using rule based reasoning. In addition to the shortcomings of this method described in that discussion, there is at least one other impediment to practical implementation. This problem stems from the fact that the input data to the rules may not be verifiable. This means that some probability or likelihood must express the data for such situations that points toward a statistical approach to knowledge based systems.[7]

The first statistical approach is Bayesian. This method begins by assigning a probability value to the conclusion of an IF-THEN rule as below:

IF: The sky is mostly cloudy
THEN: It will rain (probability = 0.6)

The Bayesian approach offers opportunities of rule based systems simply by applying the laws of probability as part of the control logic. This logic allows one to propagate the separate probabilities for each of the rules through the system to arrive at a net probability for the inference process. The two main problems with this approach are complexity and justification. The implementation of the statistical knowledge based system using classical probability involves so many more calculations that performance can be a hurdle. In addition, the complexity also diminishes the intuitive nature of the results. Many users of such systems will demand a justification of the results. A statistical approach may cloud the dominant reasons that drive the inference engine to a conclusion.

One can address the deficiencies of the statistical method by generalization. This step is called *evidence theory*. The simplest evidence relationship involves casting the premise of the IF-THEN as evidence and the THEN portion as the hypothesis. This method emerged in the 1970s as the Dempster-Schafer theory of evidence. Although this theory offers more realism, it still does not capture the process that a human expert might undertake in trading off various options to reach a conclusion.

Fuzzy logic is a method to quantify some of the characteristics involved in the decisions for a particular domain. For example, consider the possible indications of a traffic signal:

GO = { green (1.0), yellow (0.5), red (0.0)}

This approach leads naturally to set theory. The mathematics of fuzzy logic contain extensions to the usual set theoretic operations such as union, intersection, equality, and subsets. In addition, logical set operators are available as are natural language equivalents of the results.

Suppose that one considers the set of all traffic signals in a city. A subset might be the traffic signals along a particular stretch of highway. Each traffic signal has a state (green, yellow, or red) for which at any instant one can assign a numerical weight as described above. These values usually are not probabilities because the state definitions typically are not random variables. The system state is based on the condition of all of the traffic signals along this segment of the highway. That is, one can define a function that measures the strength of the relationship between the set and its members. If all of the traffic signals are green, the output will compute to 1.0 and the subset is "very" GO. If all the traffic signals are red, the output is 0.0 and the subset is "not" GO. The intermediate conditions are more likely for an actual problem and could be "extremely" or "more or less" GO. This is the *membership function*, which generalizes classical set theory for which an element is either a member or not a member of the set. In fuzzy logic, the output of the membership functions measures the strength of the relationship of an element to a set.

Figure 6-5 shows a simple example. In this figure, 11 traffic signals are separated in multiples of a length L. The timing of these signals is synchronized to update each signal every time tick equal to L/v, where v is the speed limit. Each green and red lasts five ticks, but the yellow lasts only one tick. The probability of the entire system being green is 5/11 that corresponds to "more or less GO" in the terminology discussed above. For more complex sequences, the probability density function might be time dependent.

Time	Status/Ticks →											# Green
0	G/1	G/2	G/3	G/4	G/5	Y/1	R/1	R/2	R/3	R/4	R/5	5
Δt	Y/1	G/1	G/2	G/3	G/4	R/5	G/5	R/1	R/2	R/3	R/4	5
2Δt	R/5	Y/1	G/1	G/2	G/3	R/4	G/4	G/5	R/1	R/2	R/3	5
3Δt	R/4	R/5	Y/1	G/1	G/2	R/3	G/3	G/4	G/5	R/1	R/2	5
4Δt	R/3	R/4	R/5	Y/1	G/1	R/2	G/2	G/3	G/4	G/5	R/1	5
5Δt	R/2	R/3	R/4	R/5	Y/1	R/1	G/1	G/2	G/3	G/4	G/5	5
6Δt	R/1	R/2	R/3	R/4	R/5	G/5	Y/1	G/1	G/2	G/3	G/4	5
7Δt	G/5	R/1	R/2	R/3	R/4	G/4	R/5	Y/1	G/1	G/2	G/3	5
8Δt	G/4	G/5	R/1	R/2	R/3	G/3	R/4	R/5	Y/1	G/1	G/2	5
9Δt	G/3	G/4	G/5	R/1	R/2	G/2	R/3	R/4	R/5	Y/1	G/1	5
10Δt	G/2	G/3	G/4	G/5	R/1	G/1	R/2	R/3	R/4	R/5	Y/1	5

$$G = \text{Green} \quad = 5\Delta t$$
$$Y = \text{Yellow} \quad = \Delta t$$
$$R = \text{Red} \quad = 5\Delta t$$

Figure 6-5 Traffic signal system for the fuzzy logic example.

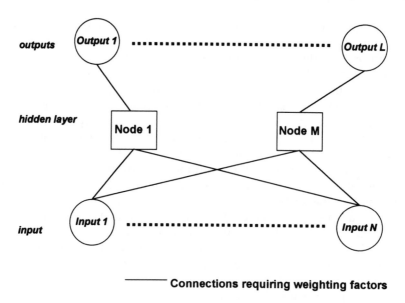

Connections requiring weighting factors

Figure 6-6   Fuzzy logic architecture.

It is possible to add feedback relationships between the output and the input stages of a knowledge based system. The strength of this feedback can be adjusted at each output cycle by matching the calculated output against test patterns for a fixed set of inputs. In this manner, the system can "learn" to solve problems.

A neural network is an implementation of such a system. The architecture consists of the three functional areas that Figure 6-6 shows. The hidden layer contains the rules, control algorithms, or other control laws dependent on the application. The input and outputs are scaled according to a node weight that adjusts the strength of the control procedure at that node. Typically, these weights are scaled to be between 0 and 1. A scaling factor of 0 turns off the control operation and a factor of 1 passes the control signal to the output stage at full strength.

The relationship of neural networks to knowledge based systems is a current research topic. However, it is certainly appealing that the neural network architecture appears so similar to the graphical summaries used to summarize a rule based system as an inference network. In addition, fuzzy methods might be added to such a system by placing another processing stage at the output of the knowledge based system to perform the test and pattern comparison for known cases. The system could feed the differences into the input stage as corrections until such corrections null out. The final step is to add statistical methods to the rule based system such as evidence theory. The resulting architecture, when properly trained, offers the possibility of producing results similar to those that an expert might formulate.

A system of this type could accept requirements and descriptions of component technologies from electronic retrieval systems. The rules of the system might represent classical trade studies that trade off factors such as

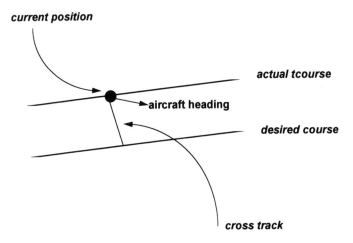

Figure 6-7 Definition of cross track.

cost against performance and schedule, cost and mission requirements against reliability, and similar factors. The outputs could be prioritized lists of interconnecting and component technologies that form a design baseline. The fuzzy control system might play such baselines against models of operational environments, manufacturing processes, repair strategies, and similar factors to achieve a low risk design. This next topic continues this theme to provide a summary of the VP approach that Part II details.

6.1.6   Fuzzy and conventional methods

This discussion provides a simple example that demonstrates some differences between conventional and fuzzy rule based reasoning. The primary cause of the differences is discrete vs. continuous output functions. Conventional rule based reasoning depends on *crisp logic* (conventional *if-thens*) whose results must be digital, taking on values of either 0 or 1. In fuzzy logic, the membership function generates outputs that can vary continuously between 0 and 1. Multiple outputs of a complex, fuzzy system can overlap to result in faster convergence and smoother operation.

Consider an autopilot that tracks a commanded course by using the *cross track* distance and rate as error signals. The cross track is simply the direction perpendicular to the desired course, as shown in Figure 6-7. In this figure, the course is a ground track while the heading is the direction in which the aircraft points. The cross track distance refers to the length of the line segment perpendicular between the actual and desired tracks. The cross track rate measures how quickly this distance changes. The rate is the aircraft velocity component (corrected for wind) in this direction. An autopilot produces a control signal that commands a heading change to null the distance error. The rate input eliminates overshoot and undershoot by acting as a damping term. This is similar in philosophy to a classical state vector approach that uses an algorithm instead of rules.

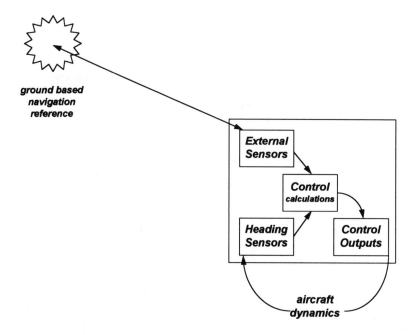

Figure 6-8   Control system block diagram.

Figure 6-8 shows a block diagram of the control system. A navigation system (such as a Kalman filter) produces the best estimate of cross track distance and rate. The basis of the estimate includes both integrated solutions from onboard, inertial sensors, and ground based references. Aircraft heading is another product of the estimator. The basis of the heading estimate might involve comparison of the inertial system result to inputs from a slaved magnetic compass, gyrocompass, or other indicators. A rule based system for this application might involve the distance and rate error inputs as shown in Table 6-1. In this table, the errors represent a variation either to the left or right of course. This deviation can increase or decrease with time. From these inputs, the autopilot can command a left, right, or no turn at the standard rate of 3 degrees per second. Table 6-1 is an *associative memory* approach because one can represent the outputs as elements of a matrix.

For a rule based system using crisp logic, the designer must establish thresholds for each of these parameters. For example, the threshold distance error might be 1 nautical mile and the rate threshold might be 10 knots.

Table 6-1   Sample Rules for the Control System

	Distance error left	Distance error right
Rate error decreasing	No turn	No turn
Rate error increasing	Right turn	Left turn

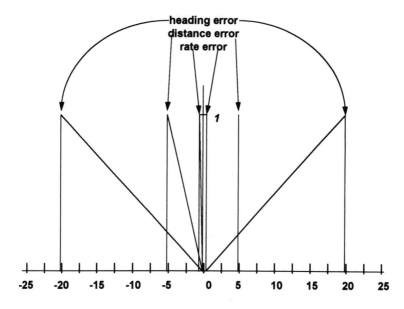

Figure 6-9 Membership functions for the signals.

Below these two thresholds, the conventional controller does not command a heading change. Above the thresholds, the autopilot might command a standard rate turn that terminates when the heading equals that of the desired course plus some standard correction. The correction usually depends upon the distance of the aircraft from the station for actual implementations, but for simplicity, ignore this refinement. In addition, if the correction involves a heading change of more than 90 degrees, then *reverse sensing* is necessary to avoid divergence.

In a crisp, rule based system, the distance and rate errors assume the value of 1 whenever the inputs exceed a threshold. However, a fuzzy, rule based system might involve the membership functions shown in Figure 6-9. The values of the outputs peak at 20 degrees for heading error, 10 knots for rate error, and 1 nautical mile for distance error.

A simulation example shows some characteristics of each rule based method. In Figures 6-10 and 6-11, the desired heading is 270 degrees. The aircraft is flying at 250 knots with an initial heading of 315 degrees. The initial distance error is 10 nautical miles and the initial rate error is 30 knots. Figure 6-10 shows the distance error vs. time for the crisp logic. Figure 6-11 shows a similar result for the fuzzy method. These graphs involve "perfect sensor" models that involve no errors in measurement.

The differences in the figures would be more dramatic if the turn rate were not limited to standard rate. Also, the error in each case can easily be driven much closer to zero by using classical *tuning methods*. These include increasing the update rate (smaller DT) or lowering the distance and rate error thresholds. The figures do not demonstrate the relationships among these variables. In the conventional rule based system, changes in the parameters of the control variables may result in performance discontinuities. The

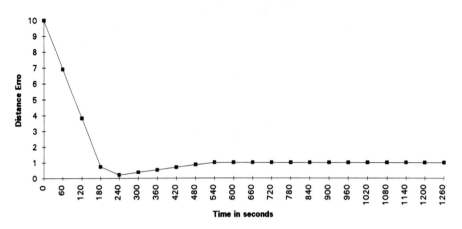

Figure 6-10   Conventional rule based control.

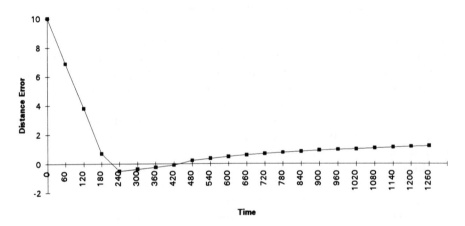

Figure 6-11   Fuzzy rule based control.

fuzzy system behaves in a smoother fashion, which may simplify the tuning process. Specifically, the fuzzy system does not oscillate so readily as the conventional controller as the rate error threshold decreases.

However, the differences in accuracy between the two methods become most apparent when the simulation includes error signals. The source of such errors might be the accuracy of the sensors and hardware control actuators. External factors that inject statistical errors could include wind shifts or turbulence. If one adds a random error to the simulation, the fuzzy controller often outperforms the conventional algorithm significantly. In addition, the advantage of the fuzzy controller is that a detailed physical model of the system is unnecessary to extract control algorithms. Instead, the fuzzy controls designer only needs a clear definition of inputs and outputs. The designer simply tunes the relationship of these parameters to achieve the correct behavior.

Fuzzy and conventional logic both have important applications. However, fuzzy logic is most valuable when clear relationships and models for the state variables may not be known. The continuous variation of the control signals eases the need to draw precise boundaries between segments of system behavior. As later chapters demonstrate, these characteristics make the fuzzy approach valuable in the implementation of expert systems for design.

6.2   VP concepts

An engineering prototype provides a vehicle for testing critical system requirements, technologies, and processes. For reasons of cost or schedule, early prototypes may not include the full functionality of the end product, but may serve to reduce risk for key design and manufacturing factors. However, successful completion of even limited testing validates both the requirements and the product baseline for release to pre-production. Therefore, the prototyping effort serves as a bridge between the design and manufacturing phases of a program.

One problem with building a prototype is that some of the manufacturing processes may not be in place before full production begins. This usually stems from the great cost in supporting production facilities so that to idle such resources is ruinous. Since a prototype is often a custom, hand-built design, it can be quite time consuming and expensive to build such units. Of course, to skip prototyping is likely to be even more expensive if problems emerge during production. Various management specialists have studied this problem and have formulated the concept of *rapid prototyping*. Rapid prototyping is an engineering capability founded on a well-defined design process, good team communication, and suitable design tools. The process concepts stem from TQM risk reduction emphasis with VP providing the design tools.

Figure 6-12 revisits an earlier discussion of the design process. Requirements generation is often the first task of the design effort. This involves a detailed technical description of the product performance in terms of documentation, such as a system specification. Initially, this specification may be rather general, but subsequent steps provide technical details. The next major task is often development and integration of analytical capability, such as simulation. Once again, the team refines simulations iteratively to achieve the proper balance between fidelity and complexity. Initially, models might capture only top level system behavior. The analytical tools provide the capability to conduct trade studies that focus on technical risk areas. This might involve selection of a major component type, such as a data processor, communication channel, or storage unit. One or more product baselines emerge from the first round of trade studies. Further testing allows the team to pare this list to a small number of candidates for which cost effective prototyping might com-

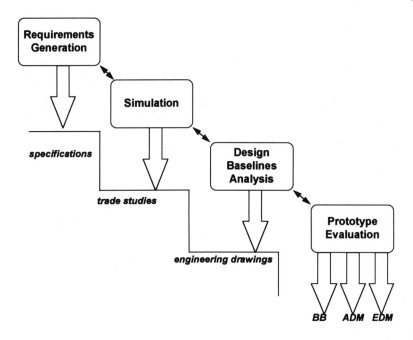

Figure 6-12   Engineering design process.

mence. Engineering drawings, such as architecture block diagrams, software documentation, circuit card layouts, or other elements capture the
features of each candidate. The earliest hardware prototype might be
designated a *brassboard* unit. This is often much larger and heavier than
the intended product, but offers the team the capability to explore interfaces and timing of the electronics. The brassboard is also an in-house
demonstration system for the system designer and is not for release to any
(even selected) customers. An engineering development model (EDM)
might be a later prototype after further requirements and trade studies.
The EDM often is the same size as the final product, but may not have full
performance or functionality. The advanced development model (ADM)
often serves as a final check before production. Often a vendor might
produce multiple ADMs to support hardware "beta" testing, initial software testing and integration, as well as any qualification testing.

The AI technologies of the last section can play an integral role in automated system design. An expert system could assist designers in producing
product baseline candidates using trade-offs among cost, technical risk, performance reserves, mean time between failures (MTBF), and similar factors.
The designer might interact with this analytical system using virtual reality
technology. This might be especially helpful in visualizing a justification of
a selection by displaying a three dimensional plot of the relationships among
the design requirements. The rules of the inference engine might be based on
fuzzy logic. The inputs might involve scanning documents or graphical
information into electronic format for which synthetic vision could be used.
Part II revisits these issues.

The notion of software prototyping is somewhat more controversial. Some managers argue that the software team does not control its top level requirements (the system integrator does) and that early prototyping is therefore impossible. This complaint is legitimate, but really is a symptom of a defective design process, as the discussion will show momentarily. Regardless, many software organizations have specialty engineers whose responsibility is product evaluation. These engineers often receive EDM hardware on which they compile and run a large library of standard software tests that explore both the functionality and performance of the candidate system. The team documents any hardware or run-time system difficulties and provides execution speed and code size information to other software engineers who are responsible for validating software requirements. The first version of the product software matures in parallel with this process so that it can serve as the next test after the standard library. Therefore, in such organizations, version one is the software prototype and is often called the "alpha" release. After the team performs some testing on the EDM and alpha system, the management may decide to provide a beta release to selected customers who also receive the ADM hardware.

Many system integrators use design methods based on SA. This is a proven method for certain types of designs but does have some drawbacks in its interaction with software designers. The SA method uses a "black box" approach that may not be compatible with the software design methods. Some designers wrap their arms so tightly around the black box concept that the connection of the hardware interfaces becomes the focus of system integration. Software may be viewed as "inside the box" and therefore not an early design concern. Unfortunately, delivery of computer hardware before software requirements are mature is one of the many pillars upon which the much touted software crisis rests.

For new systems, software may represent the bulk of the design and support costs. Therefore, this narrow focus on hardware requirements, especially hardware communications interfaces, is inappropriate. The software and systems designers must have both compatible processes and design methods. The formal methods and OOD may be the answer to this dilemma (see Chapters 2 and 3 for a description).

VP is a critical piece of this new process, as it may offer low cost, rapid prototyping. Consider a simple example. Suppose a mechanical engineer must design a lever. The early issues might include topics such as adequate strength, weight, and physical clearance to surrounding objects during operation. The engineer is likely to use a graphical computer tool to enter the drawings. Years ago, the next step could have been release of the drawings to a fabricator to build a model. However, sophisticated simulations can now perform all of the checks that such a model might provide. This allows a designer to lower cost by delaying fabrication. In addition, the approach can lower risk by providing a faster test output than the fabrication cycle. Finally, such computer based tools are often much less expensive than the iterative fabrication of multiple hardware units.

The "virtual wind tunnel" is a more complex example. This research is driven by the high cost of airfoil prototypes and wind tunnel testing in comparison to simulation. However, because of the nonlinear partial differential equations, large numbers of node points in the finite element models, and the computing overhead of visual displays, this application has required more processing power than even supercomputers could deliver. Newer computer hardware and breakthroughs in the formulation of the algorithms bring this capability within grasp. In the near future, virtual reality technology may allow aerodynamicists to interact in real time with candidates in a cost effective manner.

Many steps of the computer design process are now automated. Libraries of logic units and other specialized tools allow microcircuit designers to completely test the behavior of a baseline design before it is released to production. These behavioral models allow for testing of control, timing, functionality, and performance. A designer can use a "silicon compiler" to synthesize a physical, gate level model from the behavioral model. The gate level model executes detailed electrical analyses whose outputs include microcircuit impedances, capacitances, power dissipation, and other physical parameters. A baseline design is captured as a Netlist, which automated foundry tools use to fabricate the silicon die for the microcircuit. In recognition of this situation, many module designers now ask design tool vendors the obvious question:

If the foundry can make silicon from the Netlist, why
can't I use the Netlist in module design?

Newer design processes offer elements of this capability as the microcircuit design metaphor of Chapter 5 suggests. However, additional tool development and integration is necessary to establish a full system level design synthesis capability. Designers now build circuit card simulations from the behavioral circuit models. Usually, performance precludes complete use of physical models, but the behavioral models from which the Netlists are derived offer a powerful integration and testing environment. The physical design of the card might emerge from another design team. The tools for their work might include several discrete types of automated circuit card layout tools. These tools accept only portions of the circuit descriptions in order to establish component placements and trace routings. System simulations might draw in behavioral models for various databases and circuit cards. Automated tools can build event and message track files that emulate sensor inputs and software control. Once again, the physical design may occur in parallel using automated tools to generate engineering drawings from the component and circuit card descriptions. Therefore, many tools and processes exist now to automate the design process, but there is no fully integrated computer design synthesis capability.

Figure 6-13 shows this process. One important observation of this process is the important role of the behavioral simulations that are the basis of the entire design method. In addition, tools are available to produce

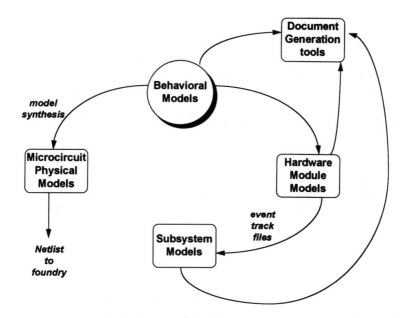

Figure 6-13 Automated computer design.

specifications from validated behavioral simulations, an approach called the "compilable specification" by some managers. This offers an important method of validating requirements, which is always a costly and difficult task.

As defined earlier, design synthesis refers to highly automated methods of producing a product baseline from input requirements. This involves much more than design tool integration, but must also include the engineering management and process control factors. AI components such as knowledge based systems can provide the framework for capturing the crucial features of successful computer engineering organizations. In this manner, a design synthesis tool can be built. Such tools are possible, but do not exist in completeness today. This is the topic of Part II.

The simulation language can have an important impact on the efficiency, performance, and testability of models. Many designers prefer VHDL (see Chapter 8), which is an IEEE standard and has wide industry support. The syntax of VHDL is similar to Ada or Pascal, but provides additional semantic checking of simulation features related to timing and interface definition.

Designers can build models using modern software engineering practices, such as OOD, even with VHDL. This method is an important step in system OOD. In addition, the integration of these simulation library units can serve as a process model for actual system integration.

The simulation, AI, and virtual reality technologies offer a powerful basis upon which to build a prototyping capability. Simulation software integration can be an important verification of the adequacy of a system design. OOD and VP play complementary roles in this method.

References and notes

1. *SCRAMNet™ Helps Bring Virtual Reality to the Financial World*, SCRAMNet™ Network Real-Time Review, December 1993, p. 3.
2. Quinnel, Richard. *Virtual Reality Enlivens High Tech Games*, EDN, December 23, 1993, p. 30.
3. Quinnel, Richard. *Software Simplifies Virtual World Design*, EDN, November 25, 1993, p. 47.
4. Barnesly, Michael. *Fractals Everywhere* (Academic Press, Boston, 1988), p. 89.
5. Gonzalez, Avelino and Dankel, Douglas. *The Engineering of Knowledge Based Systems* (Prentice Hall, Englewood Cliffs, NJ, 1993), Chapter 5.
6. Nagy, Tom et al. *Building Your First Expert System* (Ashton-Tate, Culver City, CO, 1985), p. 8.
7. Gonzalez, op. cit., Chapters 8 and 9.

Projects

Goals: These projects solidify the concepts of this chapter, as well as provide details in preparation for the design examples of the next part.

1. Identify some vendors of virtual reality simulation tools, displays, and sensor systems. Engineering magazines may help, such as given in Ref. 3 above. (You may wish to forewarn your postman of the onslaught!)
2. Identify and consider a complex system. Discuss some aspects of building and evaluating a hardware prototype that could be performed using VP instead of actual fabrication.
   (a) How many days would it require to fabricate the prototype after the team establishes a design baseline? How long to formulate a virtual prototype?
   (b) How many persons outside the design team are needed to fabricate the hardware prototype? How many for the VP?
   (c) How much equipment not related to design is necessary to fabricate the hardware prototype? How much for the VP?
   (d) Give an estimate for the turn-around time to correct problem reports for a hardware prototype (e.g., how much time is needed to clean up a hardware board design to remove cuts and jumpers?). How much for the VP?
   (e) Give an estimate for the amount of time needed to perform evaluation testing of a hardware prototype for some specific set of requirements (e.g., mechanical vibration testing). How much for the VP?
3. At some point in the engineering schedule, designers must have actual hardware. What are some trade-offs in establishing the limit point of the VP evaluation? For example, is it feasible to simulate a data processor and run actual software applications on the simulator?

4. Consider a flight simulator program that runs on a personal computer. Is this a virtual reality system? Why or why not? Refer to the three characteristics in the introduction of the chapter.

5. Discuss the need to achieve balance in the use of various sensor systems of a virtual reality application. Specifically, what might be the communication transfer rates for sound, sight, and simulated tactile systems? With which system is the application most likely to overwhelm a new user?

Part II

System design and synthesis

Figure 1 shows an overview of the automated design process that constitutes computer system VP. This model includes three major elements. The first element is the synthesis system. The product of this element is a design baseline that might be derived using ruled based algorithms for expert systems. Requirements and technology capabilities are inputs to the synthesis engine. The second element is the virtual environment, which allows the baseline to interact with operational scenarios. This element is a virtual reality technology that might include virtual manufacturing of the design, simulated product testing, and other arenas. The third element is effectiveness and sensitivity analysis. This element studies the results of the operational simulations in order to optimize selections made in formulating the baseline. The optimization might be necessary in order to:

- Change the requirements (some requirements may be unrealistic)
- Modify the rules in the inference engine (especially true for test cases).

Design synthesis is the topic of this part of the text. The subsequent chapters provide examples of some of the technologies, rules for the synthesis system, architectures of the implementation, and examples. Chapter 7 sketches the requirements process that feeds the synthesis engine. Chapter 8 illustrates some of the technology elements including object oriented simulation (Section 8.2) and automated tools (Section 8.3 and 8.4). The design rules are contained primarily in Section 9.1. The remainder of Chapters 9 and 10 offer architectural details for implementors as well as sample inputs and results of the design process. The latter might be useful as patterns that feed the effectiveness and sensitivity element in order to optimize the rule base.

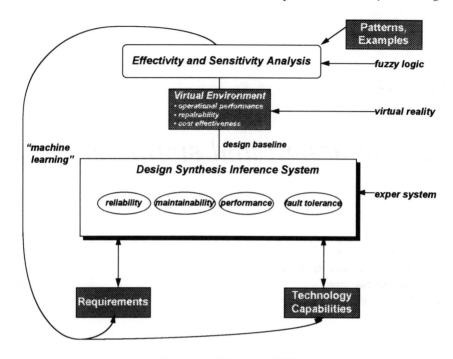

Figure 1   Elements of VP.

The presentation format of this part follows the general process model of Section 6.2. The four elements of this model are requirements, simulation, design analysis, and prototype evaluation. As previously discussed (see Figure 6-6), the design process model is more complex than might be obvious, as there is significant linkage among these elements. The OOD formalism neatly captures these relationships. This can couple with VP to form a powerful design environment in which subtle flaws are caught early using a rapid prototyping approach to risk reduction.

Although several examples illustrate the details of this process model, a single example will be the touchstone of the discussion. This example is the computer systems for a transport aircraft. This type of aircraft is familiar from normal airline operations as a pressurized, subsonic flight system for which economy of operation is an important factor.

chapter seven

Requirements

Requirements are one of the inputs to the knowledge based system that produces a design baseline. Many automated tools exist to capture and manage requirements. Therefore, electronic database formats should be readily available to feed into the knowledge based system. This chapter provides the basis for understanding the various types of requirements and their use in design.

A requirement is a technical statement of the performance, functionality, or interface definition of an element of a complex system. This chapter presents an example of requirements formulation and management based on transport aircraft needs. The significance of this topic is that other design tools, methods, or strategies rely upon careful management of product requirements in order to assess performance and risk.

Those who develop and validate requirements often speak of *tiering*. This refers to the accumulation of increasing detail as one probes more deeply into the functional decomposition of the top level requirements. Practitioners commonly describe at least three layers:

- Operational
- System
- Subsystem.

The relationship of these layers is not comparable just to the proverbial layers of the onion, but also relates to a temporal evolution. Operational requirements apply to all aspects of the product baseline and relate directly to a need statement. As such, the operational requirements must emerge very early. The system and subsystem requirements must mature in a more parallel fashion due to the greater complexity of coupling among such elements. Estimates for these quantities emerge during the design process and feed back into requirements. Although it is crucial for design integrity to assure that requirements definition is separate from design estimates, to some extent this coupling always exists. The requirements are based not just on customer needs but on what is achievable with available technology. OOD can characterize this synergism and VP can increase productivity.

7.1   Operational requirements

The operational requirements are the upper tier characteristics that determine the behavior that a typical user of the product might encounter. They are negotiated between the customer and system integrator at the time of contract award and often represent the basis of acceptance testing. For the transport aircraft example, these might include aircraft size, weight, payload, range, speed, and cost. The technical documentation for each of these areas must include an engineering statement of the requirement and the method to test it. Weight requirements provide a simple example to follow, yet offer surprisingly difficult design issues. Therefore, the subsequent discussions focus on the aircraft weight requirement. It is unimportant for this discourse to provide a weight limit. It is sufficient for this topic to understand that the upper tier requirement might be stated as:

The aircraft gross takeoff weight shall not exceed (...) pounds.

The reader should not conclude from this statement (or others) that the various upper tier requirements are independent. For example, the weight requirement might drive the designers of landing gear to rely on an exotic new metal alloy. Such a decision for the landing gear might impact cost of the project due to the newer tooling and production methods, as well as the relatively higher cost of the metal itself. In addition, this decision might have significant schedule impact. The landing gear design may be unique, which might force the fabrication of many hardware protctypes to verify mechanical properties and design features. This is a general trend of program requirements:

**Available technology in each major area drives cost
and schedule.**

In short, cost is usually not an independent variable in the design equation despite the insistence of the customer on some bottom line price. Instead, the designer must establish the functional relationship of cost to the other top level requirements as part of the creative process. Figure 7-1 illustrates these trade-offs qualitatively. Despite the complexity of many systems, the lack of systematic methods to formulate these relationships among requirements forces many designers into a trial and error process.

Trade studies are an important risk reduction component of this process. VP can speed this process and greatly reduce the cost of technical trade studies for some applications. Section 7.2 explores the formulation and execution of trade studies.

In addition to numerous technical trade studies, the managers of many programs require a cost and operational effectiveness analysis (COEA) early in the design phase. Although technical requirements are the focus of this presentation, a negative recommendation in the COEA normally represents a "show stopper" and indicates a major problem with the formulation of operational requirements. For smaller efforts, frequent communication between the customer and the development team is essential to assure a com-

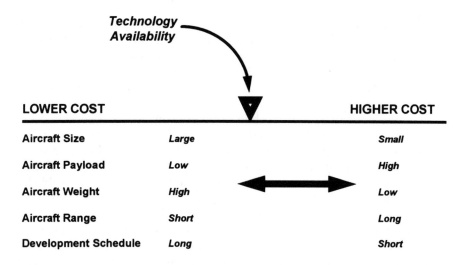

Figure 7-1 Relationships among upper tier requirements.

plete understanding of top level requirements prior to detailed engineering efforts. In a fully automated environment, the COEA might be a component of the effectiveness and sensitivity element of Figure 1 (refer to the introduction to Part II).

In implicit recognition of the links among technical requirements, the subsequent discussions do not revisit the fact that each top tier technology decision has cost and schedule impacts. Although some examples are given later, this association is generally tacit in the discussion of this (and other) texts. However, there can be other links as well. In this example, an increase in weight decreases payload, range (perhaps), and increases size. Therefore, it seems that weight is also not an independent variable. However, arguments of this type can quickly spiral out of control. An increase in aircraft payload increases size and weight, so how many independent variables are left?

Although these issues are usually far too complicated to resolve mathematically, consider a very simple model. The total aircraft weight consists of empty weight and payload, which must be smaller than some limit:

$$W = W_0 + P < L_w$$

On the other hand, the size might be calculable from some pseudo-density relationship based on historical evidence:

$$W = \rho S$$

which must also be smaller than another limit L_s. The question becomes whether the set of equations has a solution:

$$W_0 + P < L_w$$

$$\rho(W_0 + P) < L_s$$

The payload is likely to be unalterable, so this is equivalent to proving the relationship $P < L_s/\rho - W_o$ throughout the design process. Although these models are too simple for use in an actual program, they do demonstrate that an engineering group must establish the dependencies. This is commonly based on historical evidence, which can identify the relationships and sensitivities.

Finally, the real key to resolving this problem is the observation that there are several users. The air crews may have one set of preferences, the maintenance crews another, the dispatcher might have a third set, and airline management certainly has definite ideas about the aircraft design. The designer must draw out these preferences, prioritize them, and establish a balance among them. Any creative designer, such as an architect, who must balance many factors in a building design, must follow the same process in dealing with customers. A designer's unique solution for such dilemmas can bring character, a sense of personality, and greatness to the product. Regardless, these top tier requirements must be stated as clearly as possible early in the effort, as the subsequent decomposition illustrates.

7.2   System decomposition and top level requirements

System definition is an early, but significant, step in the establishment of a product baseline. Major systems are the gross structural elements of the product whose boundaries and character are easily recognizable and definable. The aircraft is the highest level of the system definition but the top tier system definitions are normally quite recognizable as well. For the transport aircraft system, some examples of such systems might be engines, landing gear, hydraulics, avionics, airframe, and electrical power generation systems. It is critical in complex designs that engineers working on different systems operate autonomously so that designs can mature in parallel to achieve the fastest time to market. An early question that the program manager must face is whether to organize around product areas or engineering disciplines. In practical terms, this is how program structure can impact the decomposition since the path of least resistance is often to organize a large program to reflect the structure of the design organization. However, many programs organize on the basis of product areas. This type of organization is the IPDT method of Chapter 2. The IPDT provides a lower risk by allowing the program manager to define engineering teams so that configuration management of interfaces and requirements is simplest and clearest.

Consider again the transport aircraft example. It is likely that a textual description of each system, its function, and the relationships among the systems is available from other programs. Figure 7-2 shows an example. This is the context, or system class, diagram for the baseline. Although this diagram is likely to appear obvious, it is the start of forming relationships among system requirements that is essential to one of the key process controls, requirements traceability.

These systems have at least three common features, namely weight, installed volume (size), and electrical power consumption. There may be many other factors. It is important to recognize these overlaps because the

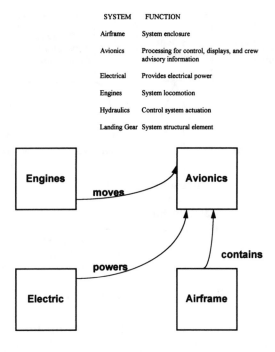

SYSTEM	FUNCTION
Airframe	System enclosure
Avionics	Processing for control, displays, and crew advisory information
Electrical	Provides electrical power
Engines	System locomotion
Hydraulics	Control system actuation
Landing Gear	System structural element

Figure 7-2   Relationships among systems.

net sum of such factors is usually the aircraft level requirement. Each system also reflects unique technical features. For example, the engines must provide a certain thrust based on requirements for aircraft weight, speed, altitude, and range. Similarly, the avionics must provide processing to support the display, control, navigation, and other aircraft and crew functions. However, even these dividing lines can be murky since, in the example just cited, the engines require digital controllers.

As the previous topic suggests, the judgment of an experienced designer is often the best guide to a strategy. In aircraft design, for example, avionics designers must fit the computing functions within the weight and size envelopes that the airframe and engine aerodynamics allow. On the other hand, avionics is a major consideration in determining the amount of electrical power needed.

Once again, the top level requirements become binding at the time of contract award. The early negotiations are quite similar to purchasing a new home, in which the contractor begins by citing previous examples to indicate the general features of the building plans. The precontract negotiations might culminate in a "red-lining" meeting in which the customer and contractor customize off-the-shelf construction blueprints. Just as with a new home, the aircraft customer is likely to be quite unhappy if the agreed upon features change very much. Although it is necessary to be flexible enough to address unforeseen events, major changes are likely to become "show stoppers" that require contract renegotiation. As a comparison, imagine the reaction of a new home buyer when the construction leader reveals that the agreed upon three bedrooms is now out of the question! The buyer of a complex system,

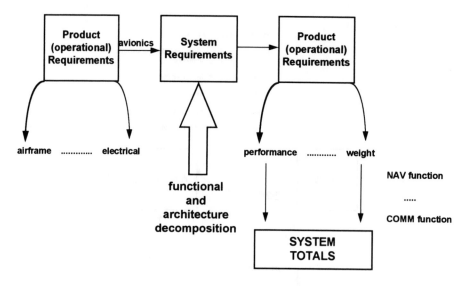

Figure 7-3 Subsystem allocations.

such as a transport aircraft, is also unlikely to be sympathetic to decreases in critical features, such as range or payload. VP can play a key role in trade studies that minimize the risk of such catastrophes by providing early, detailed insight into the relationships among operational and top tier requirements.

7.3 Lower tier requirements

The allocation of requirements is complete (but not necessarily mature) at the lowest functional level. For aircraft computer systems, this is usually a computer module or other replaceable assembly. The dividing lines among tiers often stem from an operational maintenance concept. Since avionics technicians usually only replace subsystems (boxes), maintenance hubs might repair the computers by swapping modules, and suppliers repair or replace defective modules. The same factors drive requirements in other systems. As an example, repairability of engines is another crucial factor. The maintenance approach is one of the several upper tier requirements.

The requirements allocation proceeds from top to lowest level, as Figure 7-3 shows. The system integrator often establishes the lowest tier requirements at contract award time for subcontractors. Values for top level requirements are usually based on historical data with a design margin factor.

The system (product area) for the example is avionics, which the system integrator must further decompose into functional area requirements. As mentioned earlier, the architectural implementation certainly plays a role in this decomposition. For a federated system, such as this example, each functional area maps onto a dedicated computer. This is not always the most efficient method, but provides a common enough example to warrant discussion. The subsystem requirements are composed of many areas dealing with

Table 7-1   Sample Processor Module Estimates

Item: A-1 (Processor Module Assembly)
Cognizant: John Smith
Date of report: 3/21/94

Part	Description	Vendor	Quantity	Weight (each/total)	Size/signals (each)	(each)
....						
123	Memory microcircuit	ABC	64	0.01/0.64	0.6 ×0.6	24
345	Processor	EFG	1	0.05/0.05	1.0 ×1.0	300
678	Serial controller	ABC	1	0.02/0.02	0.5 ×0.5	24
876	Interrupts controller	ABC	1	0.02/0.02	0.5 ×0.5	24
—	Printed circuit board	—	1	0.07/0.07	—	—
543	High density connector	XYZ	1	0.04/0.04	—	—
321	Transceiver	UVW	16	0.01/0.16	0.3 ×0.3	16
....						
TOTALS	1.0					
ALLOCATION	1.1					
MARGIN	0.1					

performance, functionality, and interfaces. One example is subsystem weight, which is the topic of this discussion.

The estimates for each of the areas feed into estimates for upper level tracking. As an example, consider a data processor module. The product area manager must establish a baseline tracking system that includes component quantities and descriptions similar to those shown in Table 7-1. There must also be some tracking mechanism for each component since commonly used microcircuits might appear on many different module types. Consequently, each assembly designer extracts the component data from a master library file. In addition, worksheets for the other module types are also necessary. As an example, a common modules based system might involve 10 to 15 common module types with additional application specific types.

As with all of the design areas, some sort of configuration management is essential to avoid chaos. For the data processor module design, engineering drawings are the repository of the evolving design. The area lead requires each engineer to submit master copies of the drawings and component lists on a periodic basis. The entries are also date stamped and labeled according to author. This assures that other members of the design effort have stable, validated design information from other engineers whose work is related to their own.

Configuration management at this tier of the design process can be quite expensive to a program manager due to the large volume of information and the requisite frequency of updates. However, to eliminate these controls due to cost considerations is reckless and wastes another opportunity. The same component and module databases that define the product baseline can also be the basis of simulation and prototyping. In particular, the electronic definition of the product can serve as a template that configures the intelligent VP system. As mentioned many times, a

Table 7-2   Sample Weight Requirements

System Specification =======>   (document #AV-1)
   5.1 Mission Processing

....

The mission processor shall not exceed 35 pounds in weight.

....

5.2 Navigation Processing

....

The inertial navigation processor and sensor assembly shall not exceed 45 pounds in weight.

....

· ·

Mission Processor Specification ====>   (document #AV-1.1)
   5.8 Hardware Configuration

....

The computer assembly shall include four processing module slots. Two of these slots are reserved for future growth.

....

· ·

Data Processor Module Specification ==>   (document #AV-1.1.1)
   5.11 Weight

....

The Common, 32 bit data processor shall not exceed 1.1 pounds in weight.

....

---

fully automated process can increase productivity tremendously. Therefore, the proper solution for cost concerns in this area may be more automation, not less.

7.4 Sample requirements traceability

A requirements example provides the basis for later discussions. Once again, the system is a transport aircraft. The top level requirement to trace is weight.

The requirements for contributing components are likely to be located in several different documents. As Figure 4-8 shows, the system integrator iteratively refines top level requirements down to the lowest tier allocations. If the system baseline involves a common modules approach, this may involve requirements for each computer and module type. Table 7-2 provides an example (compare this to the estimate of Table 7-1).

In Table 7-2, notice how the weight requirement flows through the document set in many ways. For example, the computer specification, designated Avionics 1.1 (AV-1.1), identifies both the module counts and overall computer weight. The module specification contains the module weight limit. Many different persons and organizations formulate these documents. The system specification is likely to be an output from the customer's engineering organization, while the other two documents emerge from various organizational elements of the system integrator. Requirements should be pushed into the lowest tier documents possible, based on the possible impact of changes to other organizations and design efforts. For example, if individual module weights appear in the system specification, not only will the customer engineering organization be overwhelmed with detail, but variations in the estimates are likely to have contractual impacts as well.

This discussion illustrates one of the many dangers of manual configuration management. A change in a requirement ripples through many requirements specifications and design databases. Unless the process is highly automated, the temptation is great to simply "complete the design and finish the documentation later." Unfortunately, the chaos resulting from this approach can ruin a program. Random chance often leads to multiple design flaws in such circumstances. Historical data proves that it is most cost effective to identify and solve design issues as soon as possible. Therefore, a highly automated approach does more than increase productivity, it can also decrease risk.

Table 7-3 shows a worksheet for the avionics weights. Since this is a federated architecture, the column entries identify both the system function and the computer name. The various module types and weight estimates are shown as row entries. In addition to the computer modules, the sum must include weight estimates for the computer enclosure, wiring harnesses, and sensor assemblies. The table entries are the number of occurrences (counts) of each assembly type. From these, an overall weight estimate emerges, shown to the right and bottom of the table. One of the several advantages to a spreadsheet of this type for cumulative requirements is that it supports *sensitivity analysis*, which determines the extent to which top level requirements estimates change as a consequence of variations in low level estimates.

As the design matures, some assemblies are likely to be over the allocated weight while others may be under the limit. If the overall computer design still meets the weight limit, the subcontractor and system integrator are likely to refine these estimates. However, if it becomes apparent that the computer design cannot meet the weight target, then contractual remediation, usually involving the customer, is needed. To avoid this scenario, cognizant engineers must maintain design margins. The margins may be very high initially, but the system integrator is likely to pressure early refinement since other system activities need design margins.

This discussion suggests several important questions about a design.

- When was the last time the design estimates were updated?
- How mature are the estimates (how much design is completed)?
- Who provided the individual module, sensor, and enclosure estimates?

Table 7-3   Weight Worksheet

Module (weight, lbs.)	Mission processor	Navigation processor	Totals (system)
Data processor (1.1)	4	1	15 (16.5)
Signal processor (1.2)	0	1	6 (7.2)
Memory (1.0)	2	1	9 (9.0)
Network I/O (0.9)	4	1	23 (20.7)
Sensor I/O (2.0)	0	2	7 (14.0)
Power supply (2.0)	2	1	25 (50.0)
Backplane (3.0)	1	1	18 (54.0)
Enclosure, harness (A/S)[a]	1 (11.0)	1 (8.0)	(183.0)
Sensors (A/S)	0	2 (21.0)	(164.0)
Current estimate:			518.4

[a]A/S means application specific.

- Who formulated the requirement for this assembly?
- How mature are the requirements (can they be validated)?
- How do these estimates compare with the requirements allocation?

Configuration management, of which requirements traceability is an element, allows a program manager a method to answer such questions periodically, such as prior to major program milestones. In addition, these issues relate directly to the risk assessment for each design area. Since risk must be quantifiable, not subjective, configuration management does more than simply bring order to the process. Instead, configuration management and requirements traceability provide the basis for tracking important risk metrics. Trade studies provide the early technical inputs for risk analysis.

7.5 Automated requirements formulation and documentation

This discussion provides an overview of the possible interaction of several automated design features. As discussed previously, tools for maintaining both requirements and baseline compliance estimates are essential.

The earlier discussion on automated tools mentioned requirements traceability. This capability is essential in any complex project because uncontrolled changes can impact other design elements unexpectedly. This observation is a key factor in the interest in *requirements reuse*. Reuse is technically very challenging, since full success requires that some technologies or design solutions not be implicit in the statement of the requirement. However, two technologies offer some encouragement in this approach.

Requirements ambiguities stemming from the imprecision of spoken language are a continual problem for complex efforts. Although numerous horror stories abound, one simple example can illustrate the pitfalls. Designer A states that at the occurrence of some event, "the signal line shall be logic one." Designer B is working on a module that senses this event based on the rising edge of a digital pulse. Designer C senses an event from a different signal line that triggers from the falling edge of the signal. Suddenly, it may not be quite so apparent what "one" means.

One solution for this problem is the concept of *compilable specifications*. This jargon means that a system designer would use a simulation language such as VHDL to describe requirements instead of spoken language. In the simple example just cited, the VHDL behavioral formulation might involve three states (on, off, and uncertain). The synthesis of the model brings in the technology basis in a natural manner that converts "on" to the requisite type of digital signal. Although this approach can solve some of the ambiguities, it certainly does not address all concerns. Nonetheless, there is much promise in the formulation of requirements, especially for computer systems, using some synthetic language that imports to a simulation environment.

Another key approach is the design method with the requirements language. If, for example, VHDL simulations are the basis of such requirements, they might be developed using OOD. This brings the discussion full circle because design of the simulations to foster their reuse leads naturally to reuse of the requirements also.

Figure 7-4 illustrates the manner in which two databases, one for requirements and the other for estimates, might interact with a TPM formulation. The requirements tracking involves identification of the source document

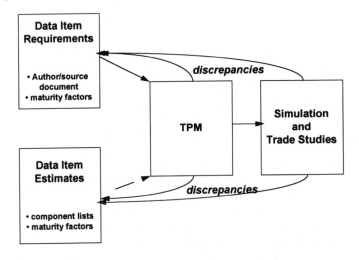

Figure 7-4 Automated requirements and estimates.

and maturity of each lower tier limit. This also involves the automated capability to trace each lower tier requirement back to first tier requirements. Estimates are another factor that the system integrator must monitor and control. The TPM involves comparing the estimate to the requirement, tempered by maturity factors. The configuration files, primarily from the estimates, could be imported into simulations to conduct trade studies. Discrepancies could be filed electronically and processed in the same system.

Return once again to the weight example. Previous discussions describe both the requirements flow down and the formulation of estimates. For a subsystem design, the baseline configuration includes module counts and types. The weight estimate for each module provides a key input to the computer weight estimate. The TPM tracks the design margin available for each module and the entire computer. Early in the development, a large margin should be maintained to avoid risk, but this can be reduced as the design matures. However, the design margin must be traded off against the cost of overbuilding the computer. The TPM, in essence, is a quantifiable measure of this trade-off.

Many of the automated tools are available now. Previous discussions have presented perhaps the most mature and powerful tools that use VHDL as a basis. Other tools include the large databases for components and estimates whose interfaces are written in standard query language (SQL). The requirements traceability tool RTM already offers an interface to SQL. Tools for STDs already provide code generators for VHDL. Synthetic requirements languages are more theoretical but probably will stem from simulation languages such as VHDL. In short, the components for a fully integrated requirements and estimation system exist but are not fully integrated. As the design environment matures, these tools offer a powerful capability for VP.

Projects

1. Obtain product literature for some type of computer system with which you have some familiarity as a user.
    (a) What are some performance requirements that must have been a consideration in this design? Give some examples of both hardware and software performance requirements.
    (b) List the hardware and software interfaces of the system. How might the requirements for these interfaces be specified?
    (c) List some functional requirements of this system. Describe the distinction between performance and function for your example.
2. Prepare a requirements flow down for electrical power dissipation. What are some of the limiting factors and linkages to other requirements areas?
3. Prepare a requirements flow down for installed volume (i.e., the space which the computer occupies). What are some of the limiting factors and linkages to other requirements areas?
4. Reconsider the electrical power dissipation example.

(a) Discuss some factors that might be the basis of a TPM for the first project (electrical power).
(b) Are these factors likely to be contained in the electronic requirements database?
(c) Is the design engineer likely to track estimates of this parameter in some automated fashion?
(d) Is simulation the basis of the estimate? If so, during which phase(s) of the design program?

5. Reconsider the installed volume example.
   (a) Discuss some factors that might be the basis of a TPM for the first project (electrical power).
   (b) Are these factors likely to be contained in the electronic requirements database?
   (c) Is the design engineer likely to track estimates of this parameter in some automated fashion?
   (d) Is simulation the basis of the estimate? If so, during which phase(s) of the design program?

6. Discuss how the configuration files that list the design baseline might be used in establishing a configuration for simulation of the prototype.

7. Consider approaches to solving requirements ambiguities.
   (a) Compare and contrast the concept of a synthetic requirements language with the natural language approach of AI.
   (b) Discuss how other AI methods, especially high resolution, stereoscopic graphics systems might also offer solutions to this problem.
   (c) Discuss the benefits and drawbacks of a software database interface, such as SQL, for tracking requirements changes and source documents.
   (d) Discuss the advantages and drawbacks of using a simulation language, such as VHDL, to solve this problem.

chapter eight

Simulation and technology capability

Rapid prototyping is an integrated approach that bridges the design and manufacturing stages of a development. Virtual prototyping is the most advanced and abstract extension since it uses virtual reality to simulate the design, assembly, or operation of a prototype system. Virtual reality methods also serve well to simulate the manufacturing processes and the operational environment. The concept of virtual reality immerses all five senses of the participant into the simulation domain. This commonly involves sophisticated visual scenes, stereo quality sound, void inputs, and in some cases even smell to simulate, for example, the smoke of a fire or some other feature.

VP relies on complex software to produce the illusion. Much of this software may be developed using software OOD and often is a *high fidelity* model, which means that it must faithfully reproduce all important behaviors and interfaces of the system. Therefore, VP can serve as a model for the system design and integration. Exactly as the simulation software is designed, tested, and integrated using OOA/OOD, the actual system can follow the same development.

As a consequence of this observation, model building becomes an important aspect. Some concepts of model building are the first topic of this section. The examples illustrate many of the relevant technologies and place them in the design context. These include object oriented C++, Ada, VHDL, and instruction set simulation.

Models that serve as faithful representations of a system may be developed using OOA and OOD. These models can serve as the basis for a description of technology capabilities. As the introduction to Part II discusses (see Figure 1), this description is an input to the design synthesis engine. Ultimately, such models may also be the basis for VP of a complex system. An output of the synthesis engine might be a simulation configuration file that draws appropriate models together for testing in the virtual environment.

This chapter includes a theoretical basis for model building that discusses issues such as the complexity of the model vs. capabilities and needs.

The second portion of the chapter includes some examples of OOD simulation models. The next topic presents elements of VHDL and the synthesis of designs. The presentation concludes with an overview of a possible integrated approach to the engineering problems.

8.1   Model types

A spectrum of model types covers computer communication mechanisms. The models differ in:

- Level of abstraction
- Level of detail
- Fidelity
- User interaction.

The level of abstraction refers to physical vs. behavioral models. A behavioral simulation could emulate the logic and sequencing for a data bus protocol, but the physical model might also account for actual signal line names and electrical operation.

To reduce complexity and improve performance, a simulation normally represents the behavior of some particular features that are relevant to a narrow segment of problems. In practice, this means that a hardware module simulation might include a software unit for each component on the module, but a system simulation is likely to incorporate only interfaces and the high level behavior of an aggregate of modules. This characteristic is *level of detail*, in which the broader the scope of a simulation in terms of module and component types, the less detail the simulation might contain about individual elements. The coverage of a general purpose simulation tends to be either broad or deep.

Level of detail is one of many strategies to render tractable simulation integration and provide adequate execution speed. A computer module might contain dozens of microcircuits with hundreds of signal line interfaces each. A system can contain dozens of module types and hundreds of modules. The testing and integration of large numbers of simulation units can rapidly become impractical. Even worse, complexity increases the probability of errors in the simulation.[1] Performance is a related issue. As the number of simulation units increases, response time slows. At some point, the delays in response become an impediment to design.

This suggests an incremental strategy to analysis. Designers thoroughly test individual components, then extract the most critical or least understood behavior. This extract becomes the basis of the hardware module simulation, after which the same process produces a subsystem simulation. A system simulation is the final step. These layers of results can be fed back to the original simulations to verify the validity of the model. In addition, as hardware prototypes become available, data from hardware measurements can also validate the simulations. In this manner, a design team can synthesize a design from requirements and technologies. Section 5.2 discusses this approach in more detail.

This discussion hints at the issue of model fidelity. No simulation is likely to capture all of the behavior of a complex system because the cost of testing and requisite performance is too high. This realization can be quite aggravating to managers who feel that they are spending huge amounts of money to achieve complete fidelity. Yet concrete examples suggest otherwise. Pilots who train on flight simulators often simply treat the devices as a different airplane type. For example, the "correct" throttle setting may produce an airspeed difference of several knots, which is large enough for a sharp pilot to sense. This, of course, does not mean that the simulator is hopelessly flawed and therefore useless. Instructor pilots are not likely to volunteer for low altitude, instrument conditions, or engine failure practice described later in this chapter. The issue, then, is how to recognize and describe the most important behavior. Primarily, this is a matter for the analysts involved in requirements analysis. Once again, Section 5.2 describes how OOD and VP can simplify this process.

Another issue is the level of user interaction. This includes at least the two features of timing and data I/O.

The simplest simulation in terms of timing is a *batch job* that the user starts, allows to run for some time (often lengthy), and then picks up the output. The other end of the spectrum is *operator in the loop*, which must function with the same timing as the actual system. In the first example, the faster the execution, the more cost effective the product. In the latter example, the simulation must replicate the timing of an actual system, sometimes at the expense of fidelity in other areas. On the other hand, some high performance computers can execute control loops faster than real time and must therefore be delayed to achieve this behavior.

Data I/O refers to keyboards, displays, or other I/O devices. As an example, a high fidelity simulator might produce impressive graphics performance for the "out the window" view. This requirement stems from a need to produce subtle visual cues from which the crew can sense ground speed, sink rate, and altitude. These displays, though, often produce the aura and enthusiasm of a video arcade.

Realism of the interaction is a hallmark of VP, of which top end flight simulators are a touchstone. Observers and participants quickly become so engrossed in these simulations that any sense of the adjacent laboratory evaporates. Imagine watching a flight crew performing an instrument approach at night. Initially, the aircraft is above the clouds. The blackness of the sky is broken by the twinkling of stars and occasional flashing lights from other distant aircraft. A pastel yellow glow flashes just to the left and a glance at the weather radar reveals that a thunderstorm will block the path to the airport within the next twenty minutes. Soon the blackness swallows the aircraft as it descends into the clouds and rain. At lower altitudes, the aircraft begins to accumulate ice and the captain nudges the throttles forward to regain airspeed lost to the drag. Streetlights and car head lamps begin to pop through the murk. The dark gray clouds turn black, then the ground appears as if someone flipped a switch. The sequenced, flashing, approach strobes point toward the runway. Ahead, the airport is brightly lit. Other aircraft crews wait patiently at the departure end of the active runway for turns at departure. Suddenly, no more than 300 feet above the ground, the crew spots a maintenance vehicle driving rapidly down the runway. The crew, already working hard, struggles to react and perform a go-around. As the first officer raises the landing gear, the aircraft shudders. The loud thumping sound and flashing red lights indicate that the right engine has lost power. The rate of climb decreases dramatically as the aircraft reenters the darkness of the clouds and rain. The aircraft accumulates more ice and the rate of climb becomes a slight descent as the stall warning system activates. The captain declares an emergency. "Request a right turn out," the captain asks, remembering the thunderstorm which is once again bouncing the aircraft. "Check fuel," he commands the first officer. "Fuel alert, five minutes," he answers.

This scenario is the essence of virtual reality, in which participants mesh seamlessly with a simulated environment. This is a powerful tool because it allows an instructor to create situations that are too dangerous to practice. However, the simulation systems needed for this type of realism are enormously expensive, in the range of millions of dollars. On the other hand, the cost of high performance computer equipment, even supercomputers, is decreasing dramatically. This trend suggests that this same technology might be appropriate for other applications, including design. Chapter 5 provides more details on the technology underpinnings of such simulation systems.

Flight simulation is a useful example because the technology is so mature. Despite the realism, it does not often fully implement AI methods, though. Imagine instead two pilots in the cockpit of an expert system that simulates the first officer. This might offer more cost effective and standardized crew training. This is also the goal of automated system design in which a customer might put complex requirements to an expert system that an operator can use to perform design evaluations.

8.2   Higher order languages and object oriented design

The purpose of a higher order language is to provide a programming interface that reflects the features of the applications instead of the underlying hardware. FORTRAN, for example, emerged in the 1960s primarily to support engineering and scientific calculations. This heritage is evident in the built-in functions and data structures of the language. Real-time programming applications usually require a blend of abstract data structures and simple access to hardware in order to provide timely and correct control inputs. Two of the more widely used HOLs for complex, real-time systems are C++ and Ada.

Many other HOLs appear in general purpose software applications. First generation simulation code is often written in FORTRAN, but this language is fading from popularity because the resulting software can be so complex and difficult to maintain. Pascal also appears occasionally, but it is not especially well suited for real-time programming.

The early days of AI also produced specialized languages that appear in various tools. The two most common special languages are LISP and PROLOG. Users of the AI tools do not usually need to know how to write code in these

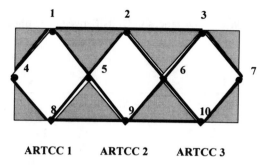

Figure 8-1   Simple model of an airspace system.

languages, but developers certainly must. In recent years, research teams also have developed AI tools in Ada and C++. The motivation for using LISP or PROLOG usually is either performance or the need to interface to some other AI tool. The performance issue, in particular, must be analyzed carefully before development. LISP and PROLOG often produce much faster AI code than general purpose HOLs. However, the speed of modern microprocessor and memory components is gradually overcoming this shortcoming.

The balance between hardware access and a general purpose programming environment might seem impossible to achieve with modern methods such as OOD that stress abstraction. However, it is also possible to build models of the underlying computer hardware using state machine theory and the concept of the virtual machine. C++ and Ada both offer support for OOD as well as direct access to hardware features.

This chapter presents a simulation example that illustrates some OOD features of C++ and Ada. The concluding remarks also describe how the virtual machine approach is the basis for using the simulation integration as a model for actual system integration.

8.2.1   Object oriented C++

This presentation considers an example based on a simple model of system capacity for the National Air Space System (NAS). The behavior is significantly simplified in comparison to the actual system so as to draw out the key features of the design approach. The presentation consists of four topics. The discussion begins with system requirements and the general design approach. The presentation concludes with a system simulation, including object oriented C++ code. This simulation demonstrates some OOD features of C++ and generally provides insight into the power of OOD.

Figure 8-1 shows the arrangement of the airports for the simplified system model. Each of the 10 dots represents an airport that is collocated with a radio navigation facility. The rectangles represent the air route traffic control centers (ARTCCs), numbered 1 to 3. The corners of the rectangles also contain navigation facilities, but no airports. Airways (solid lines in the figure) connect all the navigation facilities and no off airway routing is permitted. If the diameter of each diamond is 2 units, the matrix in Table 8-1 shows

Table 8-1   Distances in Units Between Facilities

Facility number   →
↓

	1	2	3	4	5	6	7	8	9	10
1	—	2	4	$\sqrt{2}$	$\sqrt{2}$	$2 + \sqrt{2}$	$4 + \sqrt{2}$	2	$2\sqrt{2}$	$2 + 2\sqrt{2}$
2		—	2	$2 + \sqrt{2}$	$\sqrt{2}$	$\sqrt{2}$	$2 + \sqrt{2}$	$2\sqrt{2}$	2	$2\sqrt{2}$
3			—	$4 + \sqrt{2}$	$2 + \sqrt{2}$	$\sqrt{2}$	$\sqrt{2}$	$2 + \sqrt{2}$	$2\sqrt{2}$	2
4				—	2	4	6	$\sqrt{2}$	$2 + \sqrt{2}$	$4 + \sqrt{2}$
5					—	2	4	$\sqrt{2}$	$\sqrt{2}$	$2 + \sqrt{2}$
6						—	2	$2 + \sqrt{2}$	$\sqrt{2}$	$\sqrt{2}$
7							—	$4 + \sqrt{2}$	$2 + \sqrt{2}$	$\sqrt{2}$
8								—	$2\sqrt{2}$	$2 + \sqrt{2}$
9									—	2
10										—

Note: The distance entries of the matrix are symmetric. The "—" symbol indicates an invalid combination.

the distance in units between each facility. The matrix is symmetric, which represents the fact that the distance between any two facilities is the same, regardless of which of the two is the destination.

The classes in the airspace system are:

- Hubs (collocated airport and radio navigation facility)
- Air Route Traffic Control Centers
- National Flow Control (FC)
- Users (i.e., a service request).

There are 10 objects of class hub (airports), 3 objects of class ARTCC, and 1 object of class FC. The objects are described in more detail later, but a STD for the airspace system is the starting point. The states and transitions are shown in Figure 8-2.

Figure 8-2 contains six system states, including idling. A seventh state, correcting errors, would be needed for an actual system but this presentation does not address error recovery in order to simplify the discussion. Three of the six states involve verifying the reserve capacity of system elements. Another state involves entering a request in a delay queue if the capacity of any of these resources is inadequate. The states would usually be numbered as well to establish the relationship to subdiagrams that provide more details of each state.

The model of the user is a random service request that involves a departure airport (randomly selected between 1 and 10), and an arrival airport (randomly selected between 1 and 10, but the arrival airport number must not equal the departure airport number). An identification code is assigned to each request that is valid and can be scheduled. The user also provides the number of legs between the departure and arrival airports, based on the distances shown in Table 8-1.

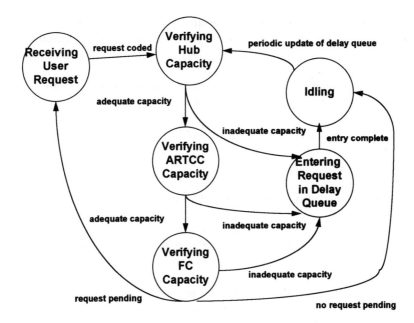

Figure 8-2 STD for the airspace example.

The simulation models hubs as separate departure and arrival queues. The departure queue is first in, first out, within some capacity determined by the maximum number of departures per hour. The arrival queue is filled in order of arrival time, up to the capacity of the arrival queue. A delay queue holds requests that cannot be serviced due to capacity limitations of the hub, ARTCC, or FC. The queues are updated and purged every 15 minutes, at which time delayed requests are reslotted.

The ARTCC model is simpler. This model contains a single queue with predefined length that represents its capacity. As an aircraft crosses the boundary between adjacent ARTCCs, a hand-off to the next controller reduces the number of users in the first ARTCC and increases the number of users in the second ARTCC by one. This queue is updated every 15 minutes.

The FC system verifies that capacity exists in the arrival hub and ARTCC systems for the service request from the user. If the resources are available, the FC provides the user with a coded identification number. This is equivalent to the transponder code that a controller assigns to a flight when the flight plan is issued. If the capacity is not available, the FC requests a delay.

This OOD system design illustrates clearly at least two of the general characteristics of modern design: abstraction and encapsulation. In software design, abstraction refers to the structure of data elements, not simply the contents. In the system context, this insight also applies to system features such as communication behavior among computing nodes. A design that requires that each ARTCC have the same communication equipment and facilities, underlies the generality of the OOD concept. The essence of encapsulation is also evident in system designs since communication protocols, performance requirements, and implementation details should be hidden except as the remainder of the system needs such insight to function. Top level system requirements such as data security and fault tolerance might be the origin of this design method.

To witness the power of OOD, consider in more detail some system requirements for the NAS design. These might include:

- Communication performance reserves
- Fault tolerance
- Information security.

The reserve requirement might state that:

> *The system shall provide a 50% I/O reserve using worst case assumptions, accounting for both hardware and software overhead.*

The system specification is likely to require many fault detection, isolation, recovery, and tolerance capabilities. The avoidance of single point failures is one example:

> *The system shall not contain any single point failures. The designer, for this purpose, shall prove using analysis or demonstrations that no single hardware or software failure shall render the system inoperable.*

Once again, the information security requirements are likely to be quite complex. One example is:

> *The system shall provide mandatory access control to assure that no software process can access information that is classified in the system security guidelines as more sensitive than the security level of that process. Specifically, no software process shall have the capability to read information from a process at a higher level of sensitivity.*

These requirements can be quite complex to satisfy simultaneously. The information security requirements dictate minimal information exchange, but fault tolerance may require significant data in order to detect and isolate faults. In addition, error recovery may require significantly more access since computer restarts and software downloads may be necessary. The information exchange associated with access control, fault detection and isolation, and error recovery also drive I/O bandwidth so that meeting the 50% reserve requirement may be challenging.

Certification is another complex consideration. Many different agencies might certify the suitability of the design for these various requirements. The agency involved in information security, for example, is likely not the same organization that tests I/O reserves.

For these reasons, historical examples are quite important. Suppose that a military aircraft system were recently certified to these same requirements. It is quite likely that the formal models, prototypes, and other risk reduction elements are still available. Even more significantly, the requirements are *validated* for this system as a result of the formal models and demonstrations. Since requirements validation is quite expensive, a natural starting point is other validated systems. Unfortunately, in some cases (such as NAS), there may not be comparable systems upon which to base such analysis.

The crossing point of software and system OOD is requirements analysis. Software designers face the same problem just elaborated, namely, how a designer can locate and reuse validated requirements from another system. Once again, the key is the formal modeling process. The level of abstraction of a system derived from SA may be inadequate to allow direct comparisons of superficially diverse system designs. OOD is formulated to simplify such comparisons through the notions of abstraction and encapsulation. This feature is a powerful aid in requirements reuse.

Many agencies insist upon evaluations that compare the new design to previously certified systems. This may include very complex, formal methods that can be very costly. OOD can be an important design approach for such systems because it captures the behavior and elements of the design in a very abstract manner. The NAS system, for example, might be based on a model for an asynchronous computer bus system, telephone system, computer network, or aircraft avionic systems. At the lowest levels, there seems to be little in common between these examples, but OOD allows a more direct comparison in terms of elements and behavior. It is the reuse of validated requirements that are based on formal models that allows maximum design reuse. Just as with software, requirements analysis and validation is a complicated and costly task. OOD specifically supports characteristics such as abstraction and encapsulation that allow designers to build such models. In a more conventional approach, it is likely that this level of abstraction would never be attained. The designers might instead leap prematurely to synthesis with many diversions regarding low level hardware operation. Performance, in particular, often diverts attention. Such a process is unlikely to easily support abstract modeling.

System design, clearly, is different than software design. However, system abstraction does require building models that can be realized as software simulations. These simulations capture essential behaviors and timings of the system in order to provide more insight to designers about both the strong and weak points of the baseline. Since this approach is so common, the discussion continues with a simple simulation of the NAS model of the last topic.

A short digression offers a demonstration of the power of automated tools. Figure 8-3 shows a subgraph of the upper right hand portion of Figure 8-2. The reset_in signal allows the user to restart the network simulation. Also, the "adequate capacity" signal is rerouted to the idling state for pur-

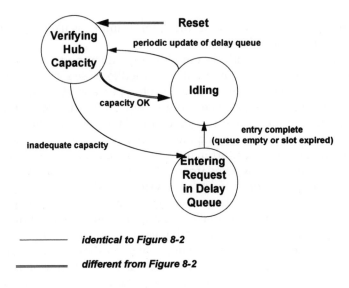

Figure 8-3   STD subgraph.

poses of this example. The legend at the bottom of the figure illustrates the differences in the STD and its subgraph.

A user can enter this STD very quickly using a graphical interface. As usual, the circles represent system states and the directed arrows indicate transitions between the states. The names of the transitions are descriptive. In addition, outputs from each state in connection with the transition (if any) are ordinarily listed below the transition name.

The user might choose to compile this subgraph into a *hardware descriptor language* (HDL). This produces a source code listing for a behavioral simulation of the graph. Figure 8-4 shows a portion of such an output. The "!"

```
state Delaying:
        IF (!reset_in & slot_expired) THEN Idling
        ELSE  IF(!reset_in & q_empty) THEN Idling
        ELSE  IF(reset_in) THEN Ver_H_Cap
        ELSE Delaying ;

state Idling:
        IF (!reset_in & update_d_q) THEN Ver_H_Cap
        ELSE  IF(reset_in) THEN Ver_H_Cap
        ELSE Idling ;

state Delaying:
        IF (!reset_in & capacity_OK) THEN Idling
        ELSE  IF(!reset_in & inadequate_cap) THEN Idling
        ELSE  IF(reset_in) THEN Ver_H_Cap
        ELSE Ver_H_Cap ;


Note: Active high transitions
```

Figure 8-4   HDL for the subgraph.

```
/*----------------------------------------------------------------*/
/*                                                                */
/*        Menu function for the simulation                        */
/*                                                                */
/*        Author:           John Newport                          */
/*        Organization:  Systems Analysis Group                   */
/*        Date:              2 June 94                            */
/*        Version:          1.0                                    */
/*        History:          Original                               */
/*        Language:        Turbo C++ for Windows                   */
/*                                                                */
/*        Behavior:        Code fragment for the HDL generation example*/
/*                                                                */
/*                                                                */
/*----------------------------------------------------------------*/

void main (void) ;

enum state      {       Delaying ,
                        Idling ,
                        Ver_H_Cap

                } ;

enum state system_state ;

enum logical    {       True ,
                        False
                } ;


enum logical    reset_in ,
                slot_expired ,
                q_empty ,
                update_d_q ,
                capacity_OK ,
                inadequate_cap ;
```

Figure 8-5   C++ for the subgraph.

symbol is a logical negation operator. In addition, the names of the states and transitions are truncated to resemble source code. Finally, the transitions are *active high* in which the asserted condition corresponds to logic value one (1).

A different tool might perform post processing of the HDL source to produce C++ code. Figure 8-5 shows an example that such a tool might generate from the source of Figure 8-4. Notice from this example also the capability to define abstract data types (such as the enumeration type) within C++.

In the process of generating the HDL source code, the compilation system performs *rules checking*, which verifies that the system can reach each state, does not become stuck at a state, and that any data transfers are properly formatted. This is a simple, but powerful AI example involving the emulation of the otherwise manual software engineering task of converting the STD into source code.

```
void main(void)

{

        if (system_state == Delaying)
        {
        if ( (reset_in == False) && (slot_expired == True))
                {system_state = Idling;}
        if ( (reset_in == False) && (q_empty == True))
                {system_state = Idling;}
        if ( reset_in ==  True)
                {system_state = Ver_H_Cap;}
        }

        if (system_state == Idling)
        {
        if ( (reset_in == False) && (update_d_q == True))
                {system_state = Ver_H_Cap;}
        if ( reset_in ==  True)
                {system_state = Ver_H_Cap;}
        }

        if (system_state == Delaying)
        {
        if ( (reset_in == False) && (capacity_OK == True))
                {system_state = Idling;}
        if ( reset_in ==  True)
                {system_state = Ver_H_Cap;}
        }


        return ;
        }
```

Figure 8-5 (continued)   C++ for the subgraph.

A simulation of the NAS system that illustrates OOD is simple to formulate. However, some assumptions must accompany the general model. Assume that a unit of distance corresponds to 60 minutes. Also, assume that each hub can service up to 1000 departures per hour, the capacity of an ARTCC is 4000 users, and the system capacity is 12,000 users.

The actual NAS design, of course, is significantly more complex than this description. However, the simulation designers can reduce complexity with a hybrid model that assumes that service requests from air carriers are scheduled, with only a small number of background asynchronous requests.[1] Such hybrid models can also be developed using OOD.

The main program of the simulation should correlate with the STD shown in Figure 8-2. This might result in the C++ listing shown in Figure 8-6. This listing clearly demonstrates the capability of C++ for object oriented programming. In this example, the classes are ARTCC and Hub. ARTCC contains 4 objects (the array center[3]), Hub contains 11 (the array airport[10]), and User contains 101 (the array airplane[100].)

```
/*-------------------------------------------------------------------*/
/*                                                                   */
/*       Main program for the National Air Space (NAS) simulation */
/*                                                                   */
/*       Author:              John Newport                        */
/*       Organization: Systems Analysis Group                     */
/*       Date:          3 December 93                             */
/*       Version:       1.0                                       */
/*       History:       Original                                  */
/*                                                                   */
/*       User Notes:                                                 */
/*       (1) an id_code of 10000 indicates an invalid departure and */
/*              arrival hub combination                           */
/*       (2) an id code of 9999 indicates that the request is delayed */
/*              this code is reset when the request is serviced   */
/*       (3) sys_max is the number of possible aircraft in the system */
/*       (4) The array size for object Aircraft should be equal to */
/*              sys_max                                           */
/*-------------------------------------------------------------------*/

# include <iostream.h>
# include <stdio.h>
# include "globals.cpp"         /* global variables */
# include "clARTCC.cpp"         /* class ARTCC */
# include "clhub.cpp"           /* class hub    */
# include "cluser.cpp"          /* class User   */

void main(void)

{

        /* Define the objects from each class */

        ARTCC               /* class name */

            Center [3] ;

        Hub         /* class name */

            Airport [10] ;
```

Figure 8-6 Main program for the NAS simulation.

The details of the functions are the next issue. Notice once again that there are three classes, the ARTCC with three objects, the Hub with 10 objects, and the User with 100 objects. The system status is updated every 15 minutes for 14 hours.

A comparison of this source code and the STD states is shown in Table 8-2. Although the airport and FC queue sizes are hard coded for simplicity, the same function could be used for all three by providing a duplicate of check_capacity() for airport and FC objects. Also, the queues for each object are not visible at the system level, only the queue depths. Therefore, the state entering request in delay queue corresponds to the code commented as "filling the delay queues" near the end of the main program. Finally,

```
        User            /* class name */

            Aircraft [100] ;


    int
                count = 0 ,
                airport_count = 10 ,
                center_count = 3 ,
                count_time = 0 ,
                count_requests = 0 ,
                index = 0 ,
                max_requests = 5 ,
                sys_max = 100 ,
                q_index ,
                q_sum ,
                av_q_size ;


    /* initialize the delay queues            */

    count = 0 ;
    do
    {
            Airport[count].Empty_DQ () ;
        count++ ;
    }
    while (count < airport_count) ;

    /* Update every 15 minutes for 14 hours */



    do
    {
                /* generate the departure requests */

                count_requests = 0 ;
```

Figure 8-6 (continued)   Main program for the NAS simulation.

idling is implicit in the code since in this formulation there are always requests.

The model also contains some functions to provide dynamic updates every 15 minutes. For example, Aircraft.User_update_position moves the aircraft in the simulation while Airport.User_arrival removes any aircraft that has reached its destination. The function Airport.Empty_DQ reschedules any delayed requests. Finally, Center.handoff removes a user from the original center queue and places it in another ARTCC queue as the aircraft crosses boundaries between ARTCC airspace.

There are some additional simplifications that may not be evident from the main program. First, ARTCC handoffs function simply as a capacity check. That is, there are no enroute delays between centers modeled. The FC capacity check functions in the same manner.

```
                        do
                        {

                                index = count_requests + sys_count ;
                                if (index < sys_max)
                        {
                                Aircraft[index].User_dep_request() ;
                                }
                                count_requests++;

                        }
                        while   (count_requests < max_requests ) ;

                        index = 0 ;
                        do
                {

                        /* update positions ----------------------------------*/
                                Aircraft[index].User_update_position();

                        /* check for arrivals ------------------------------*/
                        Aircraft[index].User_arrival();
                                index++;

                        }
                        while (index < sys_max+1) ;


                /* check the ARTCC handoffs ---------------------------------*/

                count = 0 ;
                do
                {
                        Center[count].handoff () ;
                        count++ ;
        }
                while (count < center_count) ;
```

Figure 8-6 (continued)   Main program for the NAS simulation.

The simulation might be coded to produce an output similar to the matrix shown in Table 8-3. The column index is the airport number, while the row index is the 15-minute time slot. The matrix entry is an integer that shows how many users are in the delay queue of each airport.

So that the "literati" do not become too impatient at this simple example, the reader should be aware that queuing theory can easily predict the delays without simulation. Specifically, in this case the number of entries in each queue is provided by Little's theorem, which states:

$$N = \lambda t \qquad\qquad (8.1)$$

where N =   average number in the queue,
λ =   average request arrival rate,
t =   average service time.

```
            /* check the ARTCC capacity --------------------------------*/

        count = 0 ;
        do
        {
                Center[count].check_capacity () ;
            count++ ;
        }
          while (count < center_count) ;

                /* fill the delay queues ----------------------------*/

        count = 0 ;
        do
        {

                do
                    {
                        if (Aircraft [count].id_code > 5000)
                            {
                                        q_index =
                                        Aircraft [count].departure_hub ;
                                        Airport [q_index].q_size++;
                            }
                        count++ ;
                    }
                    while (count < 100) ;

            while (count < airport_count) ;



            /* count the entries in the delay queue -----------------------*/

        count = 0 ;
        q_sum = 0 ;
        do
        {
                q_sum = q_sum + Airport[count].q_size ;
            count++ ;
        }
          while (count < airport_count) ;
          av_q_size = q_sum / airport_count ;

                /* Output the results -------------------------------*/

                printf ("%i %i\n", count_time , av_q_size) ;
                count_time++;
        }
          while   (count_time < 64) ;



        cout << "end of program" ;
```

Figure 8-6 (continued)   Main program for the NAS simulation.

Table 8-2 STD and Simulation Functions

STD state	Simulation function
Receiving user request	Aircraft.User_dep_request()
Verifying hub capacity	(hard coded queue size < 100 entries)
Verifying ARTCC capacity	Center.Check_capacity()
Verifying FC capacity	(hard coded queue size < 5000 entries)

Table 8-3 Sample Output Matrix

Hub →	1	2	3	4	5	6	7	8	9	10
Hour ↓										
1	[entries are delays in minutes]									
2										
3										
4										
5										
6										
7										
8										
9										
10										
11										
12										
13										
14										

However, the average capacity is not the issue since the requirements dictate capacity based on *worst case* scenarios. One might argue that if the average number in the queue is very small, no further checking is necessary, but the customer is unlikely to agree. One additional feature that is quite unrealistic is the assumption of uniformly distributed arrival requests over the entire 14 hours.

A final issue might be access to hardware services in order to implement the simulation. The designer may need this access to read the real-time clock, set and clear interrupts tied to the execution of software processes, or to download data to an I/O control system. Most modern computer hardware designers *memory map* such devices so that they are easy to program. This means that the software designer can access the control registers and data space of the underlying hardware simply with reads or writes to a dedicated memory location. Therefore, the HOL must support the capability to directly access a physical memory location.

Both C and C++ provide a data type called a *pointer*, which is simply the memory location of a variable. That is, one might declare an integer counter, a pointer to the counter, and then initialize the pointer to associate it with a particular counter.

```
int  count_accesses ;
int  *p_count_accesses;


..............

p_count_accesses = &count_accesses;
```

The variable count_accesses will contain the current value of this counter. After assigning the pointer to this counter (the last statement of the code fragment), the pointer indicates the physical address of this variable in memory.[2]

In C++, one can dynamically (during program execution) allocate or free memory for such devices using the predefined functions malloc() and free().[3] Another important feature is the capability to declare such variables as volatile, which prevents highly optimizing compilers to eliminate such data structures. This can happen when the optimizer does not sense the complex relationships among the data structures properly and views such declarations as unreachable code.

One final issue is type conversion. Many times, a data word can arrive from the controller in a data type that does not match the corresponding type in the application program. A typical situation involves transfer of an integer from the controller to memory, but to an application that must use this data as real numbers. Type conversion is the mechanism by which the HOL supports this translation. C++ allows the programmer to perform type conversion simply with assignments statements (i.e., equate the integer variable to a temporary real variable). This can be quite dangerous with real-time programs, though, in the hands of an inexperienced programmer. The next topic revisits this issue.

This chapter presents a simulation example in C++. The example illustrates the OOD class and object characteristics that are provided with the compilation system. The discussion also provides a description of the memory interface that a programmer might use to access memory mapped hardware. C++ offers some built-in functions that allow the programmer to control memory during execution. The next topic revisits these issues using Ada.

8.2.2   Object oriented Ada

The previous topic provided an example of a simulation in C++. This topic presents some features of the Ada HOL as background for subsequent examples.

Ada is a HOL developed by the Department of Defense (DoD). The purpose of this development was to simplify embedded software support by moving to a common HOL. In addition, the design of the HOL was to incorporate extensive error checking to assist in catching errors early in the development process. These features also simplify the maintenance of complex software. The efforts culminated in the Ada Language Standard, MIL-STD-1815A, in 1983. Shortly thereafter, the language was adopted as a standard by the American National Standards Institute.[4] In 1994, a major update was released that is designated Ada-9X.

The use of Ada has slowly spread outside the DoD to other developers of complex software. One such application is simulation, especially flight simulation software. The motivation for use of Ada in such applications is usually an attempt to decrease maintenance costs.

Ada code is quite similar in appearance to Pascal. However, this appearance is misleading because Ada provides several features that Pascal does not offer. The primary distinctions are rules regarding the data types and process scheduling support. The support for data types include a richer capability for the user to define types. In addition, Ada offers an "access type" that is useful in real-time code such as hardware drivers for memory mapped devices. The scheduling support includes a prioritized tasking model.

Just as with Pascal, the basic structural unit of Ada is the procedure. However, Ada also provides the capability to assemble many related procedures into a package. A package is broken into a specification that contains the interface and a body.

Finally, Ada fully supports polymorphism, includes inheritance rules for data types and subtypes, and supports many of the other advanced software features discussed in Chapter 3. For this reason, it is necessary to provide defaults for environment variables at start-up. Ada includes most such information in PACKAGE SYSTEM which is required for all implementations. SYSTEM includes definitions for the various data types such as float, integer, and others. These definitions include the range of the variables that may differ among machine types due to the precision of the computers. SYSTEM also includes operator definitions for these data types such as the arithmetic operators for integer and float types. Finally, SYSTEM contains some constants that define the update rate and precision of the computer's clock.

Ada also provides several packages for I/O. PACKAGE TEXT_IO is a terminal interface that includes procedures to write to the computer screen or read from the keyboard. There are also packages for the software developer to perform I/O with float or integer data types. Finally, the software developer can rename these packages to provide an interface for application specific I/O.

Generic procedures are an Ada construct that a designer can use to implement OOD. A generic subprogram (procedure) is a template that can be adapted to various applications by an instantiation.[4] That is, the generic subprogram offers an interface definition that one can shape for various applications. The Ada language system offers many such generics for I/O. An example of instantiating a new procedure that occurs frequently is floating point reads and writes to the computer terminal:

package REAL_IO is new FLOAT_IO(float);.

As suggested in the previous discussion, real-time programmers often need a method to place data at a physical memory location. The Ada constructs are conceptually similar to those of C++, although the language syntax is different. PACKAGE SYSTEM contains various constants that define the word size and possible range of addresses for a particular hardware implementation. This package also contains a type declaration called ADDRESS. An implementation for a variable called CONTROL_REGISTER_LOCATION might appear as below:

CONTROL_REGISTER_LOCATION : SYSTEM.ADDRESS ;

....

for CONTROL_REGISTER_LOCATION use at 16#1234# ;

which indicates that the software can access this control register through memory location 1234 hexadecimal.

Ada also provides *pragmas* for various memory operations. A pragma is a command line entry for the compilation system. PRAGMA VOLATILE(variable) functions in a manner similar to the C++ construct to force a variable to be stored globally. PRAGMA PACK(variable) instructs the compiler and linker to eliminate gaps in mapping records and arrays for the most optimal use of memory. Final, PRAGMA SHARED(variable) forces tasks to synchronize their updates to the given variable to assure data integrity in the system. However, PRAGMA SHARED can produce unexpected results, so many software organizations do not allow programmers to use this construct.

Type conversion is necessary in Ada for the same reasons as in C++. However, the designers of Ada found that unintended type conversions were the source of many programming errors, so to increase its testability the language enforces *strong type checking*. This means that a statement (such as an assignment) that contains variables of different types will not compile successfully. This feature can be quite frustrating for new Ada programmers but does lead to more fault tolerant code. However, the designers also recognized the need for type conversion and provided a generic function called UNCHECKED_CONVERSION(variable). This function turns off type checking for the variable listed. Of course, great care should be taken in using this function as it eliminates certain Ada safety features.

The relative advantages of Ada vs. C++ are the subject of endless debates. C++ has the advantage of wider usage, the significance of which is that more trained programmers and less expensive development tools are available. However, for very complex software, especially if fault tolerance or long-term maintenance are design drivers, many customers insist on Ada. Regardless, either can serve as the HOL for a simulation development. In this respect, they are general purpose languages. There are also special purpose languages such as VHDL, which is the next topic.

8.3   Elements of VHDL

IEEE-1076 describes the technical aspects of the VHDL language. First time VHDL users with Ada programming experience will recognize the syntax, but there are some differences. The distinctions are probably equivalent to shifting between Pascal and Ada, for example. However, there are some important software environment characteristics.

One aspect is the addition of several new data types in package STANDARD (similar to the Ada package SYSTEM). One of the most fundamental is the type time, whose range is implementation dependent, but the interface for which defines time ticks in increments of femtoseconds. Also, there are other packages that define the technology. In circuit design, for example, the package is often called STANDARD_LOGIC and contains essential timing limits, electrical characteristics, and some elemental data structures and support packages from which a designer might assemble complex logic elements. The STANDARD_LOGIC data structures usually contain a data type called signal. In the case of TTL, the timing parameters of the signals stem from the electrical characteristics of the technology, such as submicron complementary metal oxide semiconductor (CMOS). This data type also includes a definition of the possible states of a signal. The number of states is often quite large, as the third element of a tri-state system (on, off, uncertain) usually contains multiple values related to both transitions and errors, such as rising, falling, or out of tolerance.

Another package is TEXTIO (versus the Ada standard package TEXT_IO). Unlike Ada, TEXTIO provides a read and a write function for each data type that STANDARD contains. A microcircuit example illustrates some of these features.

Figure 8-7 shows a row-column address selection application specific integrated circuit. The memory devices, shown as the rectangles in the diagram, provide a 4 bit data path and a 10 bit address. A write enable input determines whether the access is an input or output request, while the chip select enables the memory device. The diagram shows that the rows break the 32 bit data word into 8 rows of 4 bits each, while a bank select line enables a column of 8 memory devices by turning on the chip select line for each.

A VHDL code fragment is shown below the diagram. The package M_BUS defines the states, transitions, and error conditions for the on module memory bus. In addition, this package must contain the data structures that correspond to address, data, and control lines. Ordinarily, the data processor is the master of the M_BUS and configures these parameters in a package for the processor model. This fragment demonstrates address selection logic and signals error conditions, if necessary. Finally, the access time is shown as the "after" statements.

Previous discussions present the distinction between a behavioral and a gate level model. The behavioral model captures essential timing, state, and transition logic, but the physical model adds implementation details. The automated synthesis function might be the method of choice for a microcircuit designer to move from the behavioral to a gate level (physical) model.

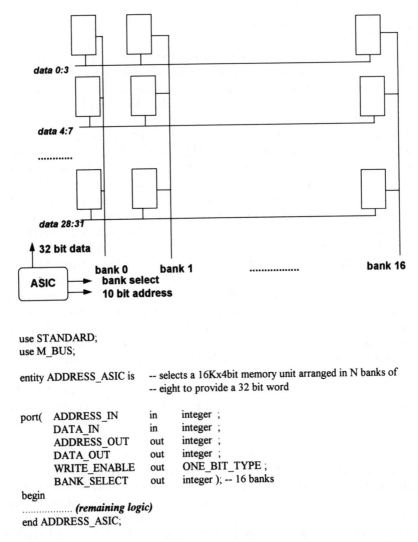

```
use STANDARD;
use M_BUS;

entity ADDRESS_ASIC is    -- selects a 16Kx4bit memory unit arranged in N banks of
                          -- eight to provide a 32 bit word

port(   ADDRESS_IN        in     integer ;
        DATA_IN           in     integer ;
        ADDRESS_OUT       out    integer ;
        DATA_OUT          out    integer ;
        WRITE_ENABLE      out    ONE_BIT_TYPE ;
        BANK_SELECT       out    integer ); -- 16 banks
begin
............... (remaining logic)
end ADDRESS_ASIC;
```

Figure 8-7   Sample VHDL code fragment for a memory select ASIC.

VHDL is also often used for parallel bus and module simulations. A synthesis function at this level is not too difficult to imagine, but obviously very resource intensive to implement.

Consider the synthesis of a module design. Suppose that both behavioral and gate level simulations of all the microcircuits run successfully. The next step involves definition of the module layout. This involves placement of the microcircuits and the traces that connect them together. The software that performs this function is effectively a finite element simulation that involves a three dimensional matrix. The signal lines of the microcircuits connect to certain node points in this mesh, as Figure 8-8 shows. As a simple example, suppose that layer 3 is Vcc, the power source. If pin 1 on the two microcircuits in the figure must connect to Vcc, their traces might be as shown in the bottom of the figure. Of course, the completed design

```
architecture BEHAVIOR of ADDRESS_ASIC is

        TEMP_BS ,
        TEMP_AO ,
        TEMP_DO : integer ;
        TEMP_WE : ONE_BIT_TYPE ;

begin

        TEMP_BS := -1;        -- begin column select
        if     (ADDRESS_IN < 16*1024) then
                        := 0;
        elseif (ADDRESS_IN < 32*1024) then
                        TEMP_BS := 1;
........... (remaining logic)
        elseif (ADDRESS_IN < 256*1024) then
                        TEMP_BS := 15;
        end if;

        TEMP_AO := ADDRESS_IN - TEMP_BS * 16 * 1024 ;
        if (TEMP_BS = -1) then
                MBUS_ERROR = true;
        end if;

        TEMP_WE := 0 ;
        if (M_BUS_STATE = WRITE) then
                TEMP_WE = 1 ;
        end if;

        TEMP_DO := DATA_IN ;    -- four bits per row in hardware

        BANK_SELECT     <= TEMP_BS        after 10 ns;
        ADDRESS_OUT     <= TEMP_AO        after 10 ns;
        DATA_OUT        <= TEMP_DO        after 10 ns;
        WRITE_ENABLE    <= TEMP_WE        after 10 ns;

end BEHAVIOR;
```

Figure 8-7 (continued)   Sample VHDL code fragment for a memory select ASIC.

must provide signal lines for each pin on the microcircuit. The automatic routing software becomes somewhat of an expert system at this point because intelligent placement is necessary to avoid clumsy designs. Specifically, digital signal line trace lengths must be minimized. The need for automation is apparent with the observation that a typical module design might contain 64 (or more) memory circuits, plus numerous other control and support logic circuits.

Return once again to the behavioral simulation of the module. The routing procedure is the critical first step in the complete simulation of a module since, among other factors, signal propagation times are influenced by path lengths. The routing software produces the physical configuration, but a fully automated synthesis operation is unavailable. Nonetheless, the elements of a completely automated synthesis capability exist and are likely to be more fully integrated by software vendors in the near future. This might elevate module design and simulation to the same level as microcircuit

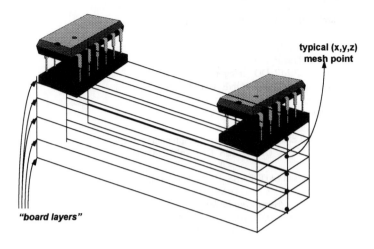

Typical traces to Vcc:

Circuit 1, pin 1 = { (10,15,1), (10, 15, 2), (10, 15, 3)}
Circuit 2, pin 1 = { (30,15,1), (30, 15, 2), (30, 15, 3)}

Figure 8-8   Typical module mesh points.

design. The current limitations are primarily cost effective, high performance computers.

Once models are available for hardware modules, a complete computer system simulation is possible. Many design shops perform this task now for unconventional system baselines as a risk reduction method. However, it is impractical to simply link together the various module and bus simulations because of the computing resources necessary to execute such a simulation. Suppose, for example, that a system consists of node points separated by 0.2 in. with 7 layer boards of 8 in. square, then a single module requires 112,000 node points, while a system of 5 modules requires 560,000 nodes. These calculations are similar in complexity and design to the "virtual wind tunnel" example mentioned in Part I. A massively parallel supercomputer is a prerequisite for this type of simulation so that results are timely. As technology and architecture breakthroughs decrease cost per system performance, complete system simulation will emerge.

This discussion concludes with a sample behavioral element for a data processor module simulation. The operation is the start-up procedure. Two types can occur, a cold start or a warm start. The cold start involves resting of all microcircuits, execution of the initial diagnostics, and loading of low level kernel code needed to begin the execution of an application program. A warm start effectively is an operating system reset since it bypasses diagnostics and simply clears the processor registers and reloads the kernel.

Figure 8-9 shows a sample logic diagram for a warm start, while Figure 8-10 shows the timing. The logic diagram might be the detailed drawing of

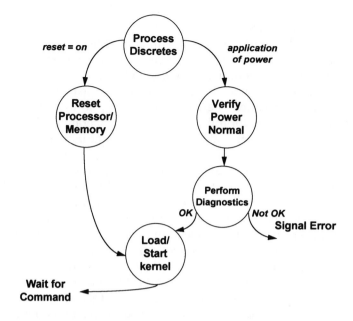

Figure 8-9 Warm start logic sequence.

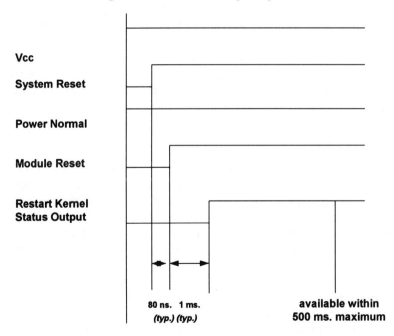

Figure 8-10 Warm start timing diagram.

the overall module STD. The substates support the top level state "initializing." This illustrates that the top level states of the STD might require many sublevels to fully capture the states and transition logic. The timing for the start-up sequences is based on the Vcc and Power Normal indications that the power supply provides. For a warm start, these two power signals remain on. Instead, the trigger for a warm start is the reset discrete, not a power cycle. The system reset discrete sets the module reset line, which connects to the reset input of the processor and memory microcircuits. The processor microcircuits usually have a reset pin that clears the registers and initializes its communication interface. The memory address ASIC (see previous example) might perform the memory clear operation. There is some small time delay between the system reset valid condition and the module reset valid due to path length delays. After reset, the processor begins execution at address 0 since everything is cleared. The instruction at address 0 is a jump to the entry point of the start-up code in ROM, also known as the boot ROM. The boot ROM fetches the monitor code from another ROM location and loads it into memory. After verifying that the kernel load is successful, often using checksums, the boot code terminates with a jump to the starting location of the kernel code. This completes the start-up.

This discussion brings the example full circle to the concept of synthesis. Automatic code generation was a topic in Part I. One of the tools that this presentation cites can generate VHDL code from STDs and timing diagrams. This initialization example shows how powerful such a tool might be. Most designers might prefer not to spend large amounts of design time with the start sequence, but clearly if this sequence does not function properly, the remainder of the design effort is useless. However, hand coding VHDL with all of the proper characteristics offers many opportunities for error. Multiple software engineers might enter portions of this code from hard copies of design specifications. The ambiguities and oversights of these documents can easily cause subtle, but fatal errors. The graphical methods are much more intuitive and less likely to introduce error.

8.4   Instruction set simulation

The alert reader may have noticed that the previous discussions focused on hardware design. However, a fully integrated process must allow software designers to participate and interact with other members of the design team using common tools. Once again, the computing resources necessary to establish this environment are enormous, but the individual technologies exist now. High performance parallel computers may be the breakthrough that makes this capability affordable.

An instruction set simulator suggests some general features of the software technologies needed for integrated design. The software engineer designs, codes, compiles, and links a program as if it is to run on the actual hardware. The simulator can interpret the output file and translate its machine instructions into the native operations of the host computer.

Consider some simple examples, such as the load and store instructions that are so prevalent among modern data processors. The simulator code might contain large arrays that simulate memory. Beyond a certain size limitation, these data structures might be augmented with storage into open files on the hard disk. The simulator must also contain data structures to emulate the internal registers of the processor. Therefore, a store instruction might appear to the simulator input as:

STORE R1, R2,

which means store the contents of user register 1 at the address contained in register 2. The simulator must calculate the location of the address on the host system. This address might be located on the system hard disk drive, in memory, or on some other storage device. Storing the value is not difficult, obviously, once the address is translated into a storage location on the host system.

The simulator allows software engineers to execute code before hardware is available, but typical implementations have many drawbacks. First, it is usually only possible to simulate a single processor. As a result, the simulator cannot reproduce any interactions among hardware modules, such as communication events. Another common problem is that the simulator usually does not include interrupts, or other low level hardware events that the run time system (RTS) must service. This often includes essential omissions, such as the real-time clock interface. In order to overcome these deficiencies, the software engineer must provide simulated services within the application code designed to run on the simulator. Unfortunately, these limitations often relegate simulators to a discard item as soon as actual hardware is available.

Consider instead an instruction simulator interacting with a hardware module simulation. The simulations of the various microcircuits can provide the support logic necessary to overcome the deficiencies just cited. However, it would be quite inefficient to simulate the entire module. Instead, *selective linking* might draw in only those microcircuit services necessary to support the RTS features for a particular application. Nonetheless, performance of the host is likely to be a problem, which suggests that separate computing nodes support the low level hardware and instruction set simulators.

The next capability is a multiple module environment. The addition of more computing host nodes allows the capability. The picture that emerges, then, is a multicomputing system such as the one shown in Figure 8-11. A massively parallel host system might be necessary to provide the performance necessary for timely response. Nonetheless, the system is likely to run much more slowly than real time.

A simple calculation illustrates the origin of the performance penalties. Procedure call overheads may be as high as 100 µs. If 10 such calls per simulated instruction are necessary on average, then 1 ms per instruction is needed. Actual RISC processor hardware may execute "core" instructions in one clock cycle or fewer, based on pipelining and other low level hardware allocation. At 25 MHz, one clock cycle is 40 ns. This means that the simulated instruction runs 25,000 times more slowly than actual hardware. Therefore, simulating 1 s of software execution requires 6.9 hours for this simple case. Highly parallel simulation can make this result much more reasonable, but the simulated code is likely to require tremendous host computer resources.

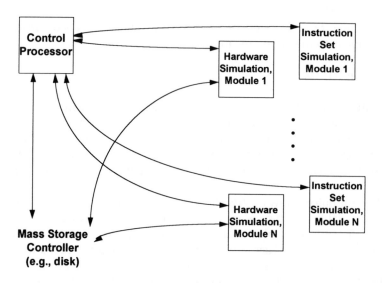

Figure 8-11   System simulation environment.

A final consideration is the cost. A parallel supercomputer might cost $1.5 million. If the simulations are used for many system designs, the supercomputer cost might be defrayed by early detection of design flaws that might be quite expensive to find later in the process. As with any expensive equipment, continuous usage is the key to cost effective operation. Therefore, while every design team may not find it cost effective to have such simulation resources, it is quite likely that several specialty vendors will emerge to support this design activity. The key elements in determining whether large scale simulation, or VP in general, is cost effective may be economy of scale and reusability of models and results. OOD is a method that can support both.

8.5   Significance of the simulation tools

This chapter presents some code fragments that bring some insight into object oriented features of HOLs and some software development tools. These discussions illustrate several important features about the software capabilities necessary for design synthesis.

First, it should be apparent that the synthesis capability is not dependent on the HOL. That is, it is equally possible to build a synthesis engine from C++, VHDL, or Ada models. The synthesis capability involves transforming a behavioral description into a physical design. This requires that the simulation environment provide numerous support packages. In the case of microcircuits, these are the cell libraries that provide gate level descriptions of primitive logic for a given silicon process. For system design, this support might include performance models and hardware component indices for parallel and serial buses, various kinds of processors, memory elements, and other hardware building blocks.

The simulations provide the capability to extract relevant behavioral and performance information about the technologies that a specific simulation models. These elements are the technology capability descriptions shown in the introduction to Part II (Figure 1). The next chapter looks at some rules that might be the basis of formulating a design baseline.

References and notes

1. Yourdan, Edward. *Structured Design* (Prentice Hall, Englewood Cliffs, NJ, 1979), pp. 68–73.
2. Perry, Greg. *C By Example* (Que, Carmel, IN, 1992), pp. 533–536.
3. Schildt, Herbert. *Using Turbo C++* (Osborne McGraw-Hill, Berkeley, CA, 1990), p. 340.
4. Cohen, Norman. *Ada as a Second Language* (McGraw-Hill, New York, 1986).
5. The simulations of this chapter were written by the author.

Projects

1. Discuss the features of C++ that the language designers provide to support OOD. How do these features compare with those of other languages? Does VHDL have OOD support? Hint: There are many papers available from technical journals on this topic.
2. Based on the discussion of instruction set simulators, consider a common low level software interface.
   (a) Is it practical to have a common data processor instruction set for all machines? Discuss some approaches to this problem that might include:
   - Instruction translation (conversion of common assembly code into native code at compilation time)
   - Instruction emulation (real-time translation of common assembler code into native code). Once again, the literature offers many papers and articles on this topic.
   (b) Discuss why a common assembly language might be beneficial. Specifically, examine how this approach might speed simulation development, especially in VHDL.
3. Describe some general areas of theory that represent models for components of computer systems.
   (a) What is automata theory? How does this theory relate to STDs and timing diagrams? Does this theory relate to a design method or is it independent of the design method? Of the three requirements types (functional, interface, performance), to which type does this theory most closely relate?
   (b) What is queuing theory? To which type of requirements does this theory most closely relate? What are some limitations of this theory?
   (c) What is scheduling theory? Hint: Consult a text on operating systems.

(d) Use these observations to discuss the model elements that are necessary to predict the performance of a computer network. Discuss the trade-off between complexity and completeness. For example, should software units model all aspects of hardware and software in the simulation?

4. Consider the simulation of a processor and its memory system.

   (a) What are the advantages of using gate level simulations of the individual microcircuits? What are the disadvantages?

   (b) Discuss how performance estimates can be formulated. For example, is it possible to estimate the number of clock cycles needed for each assembly language instruction? What factors might affect the accuracy of this estimate? How can this be the basis for estimating the performance of an application?

   (c) Suppose that the processor is a new design. Is it possible that any previous simulation software might be reusable? Give a simple example. Could OOD be a factor in such reuse?

5. Obtain the specifications that contain the requirements for some particular data bus. Suppose that a customer has just required these specifications as part of a risk reduction "quick look" at the bus behavior and performance. Your job is to design a simulation system to address several important issues.

   (a) How might you formulate a measurement of the end to end message latency? This is the amount of time required from "interrupt to interrupt" — a software interrupt on the source node initiates the transfer and an interrupt on the destination node signals the completion of the transfer. The customer is interested in worst case, best case, and average timings.

   (b) Does the specification contain a STD and a timing diagram? Describe how these can simplify the simulation design.

   (c) Does the specification include electrical characteristics? Describe a method to use simulation to establish electrical transfer limitations. Discuss how the results of these simulations might be incorporated into the bus performance simulation to add realism and accuracy.

   (d) Fault tolerance is an important concern of the customer. In terms of bus protocol operation (behavior), describe some symptoms of possible failure modes. How might you use simulation to explore these failure modes?

chapter nine

Design capture

MIL-STD-499B describes design synthesis as tasks through which "the performing activity shall *define and design solutions* for each logical set of functional and performance requirements in the functional architecture *and integrate them* as a physical architecture" (emphasis added).[1] The output is a design baseline that consists of detailed technical descriptions such as engineering drawings, component lists, process definitions, and test plans. Program managers often undertake trades studies and prototyping efforts with such baselines prior to production in order to reduce the chance of undetected errors.

In the VP realm, baseline analysis might be so highly automated as to represent an expert system. Although such a suggestion might conjure images of computers designing themselves, this design flow can speed the analysis to lower cost and risk. The VP elements serve as a bridge between design and manufacturing activities. This chapter provides an overview of the interfaces and analytical rules for such an expert system.

The discussion of expert systems in Section 5.1.3 highlighted some key features. The characteristics of the user interface are critical to software efficiency and accessibility by computer novices. This interface might be based on a natural language approach, a graphical user interface, or some other presentation format that is simple and fast. The analytical rules must be simple enough to provide rapid software execution yet complex enough to cover the subtleties that an experienced computer designer might recognize.

The overview discusses two general classes of problem domains. The first involves a new system design while the second consists of a software architecture reformulation on an existing network. The chapter concludes with an example of each.

9.1 Expert system for design

This topic addresses the rules that might form the basis of an expert system for embedded computer design. The discussion relies heavily on the requirements and simulation elements, examples of which the previous two chapters present. The rules help to establish an engineering baseline from which

```
main function 1                1  100      (name, MIPS, size)
main function 2                2  200
main function 3                3  300
main function 4                4  400
main function 5                5  500
20                                     rate
     0  32  16   8  64                 ⇒        source
     8   0  16   8  32                 ⇓        destination
    16  16   0  16  16
    32   8  64   0   8
    64  32  16   8   0
10                                     rate
     0   0   2   4   0                 ⇒        source
     2   0   0   0   0                 ⇓        destination
     4   0   0   0   0
     0   0   0   0   0
     0   0   0   0   0
```

Figure 9-1 User input functions.

the VP and ADM emerge. The discussion consists of three elements, the user inputs, analysis, and analysis outputs. The user inputs are typical questions and data that requirements, hardware, or software engineers might formulate. The analysis simulates the design process that an experienced computer system designer might follow. The output contains configuration files that an automated design tool might use to set up and run behavioral simulations of the system baselines.

9.1.1   User inputs

The user must characterize the requirements of the intended applications. This includes the name of each main system function, performance requirements for processing and I/O, storage size, and physical characteristics. As discussed in the introduction to Part II (Figure 1), requirements and technology capabilities are the two inputs to the design engine.

The functional requirements that drive the analysis might be contained in an electronic database or requirements tracking tool. Figure 9-1 provides an example of such characteristics that the expert system can feed to the analysis section. The first entry consists of a function name, processor throughput requirement in millions of instructions per second (MIPS), and the storage size needed in KB. There are five functions listed in the figure.

It is also necessary to characterize the information flow in the system. Figure 9-1 shows an example. The numbers below the function names in Figure 9-1 are the rates and data exchanges among the functions contained in the square 5×5 matrices. The first rate is 20 Hz, followed by the transfer matrix. The first matrix contains the number of bytes transferred among functions with the row as the source and the column as the destination function in the order the names are listed above this in the figure. For example, function 2 transfers 32 bytes to function 5, function 4 transfers 64 bytes to function 3, and so on. The second matrix contains the transfers at 10 Hz. The analysis section transforms these inputs into net

What is the name of this system?	*Test_system_1*
What is the name of the function?	*main function 1*
How many MIPS does this require?	*1*
How much storage (KB)?	*100*
Do you wish to enter the I/O?	*No*
Do you wish to enter another function?	*Yes*
What is the name of the function?	*main function 2*
How many MIPS does this require?	*2*
How much storage (KB)?	*200*
Do you wish to enter the I/O?	*No*
Do you wish to enter another function?	*Yes*

·········

Do you wish to enter the I/O?	*Yes*
How many bytes does main function 1 send to main function 2?	*32*
What is the slowest rate for this transfer (Hz)?	*20*

·········

User Inputs

Figure 9-2   System function preprocessor.

transfer requirements and verifies that the system can deliver the user requested performance.

The information contained in Figure 9-1 is likely to be produced by a preprocessor that queries the user to build this figure. The preprocessor could ask questions similar to spoken language. Figure 9-2 shows an example session. The processed inputs might then be dumped to a data file for replays in the analysis section. The system name, entered by the user, would be the file name. These inputs might also be derived directly from an electronic version of the system specification by a program that isolates and tracks requirements.

Technology capabilities are the second primary input. The analysis section requires a database of hardware module types and capabilities. This includes core processing module types such as data processor, signal processor, memory, and system I/O. Figure 9-3 shows some sample menus for building these descriptions. The discussion of analysis describes the meaning of these parameters. This database is likely to be built by hardware vendors or systems engineers involved in market surveys and technology studies.

9.1.2   Analysis

The analysis rules must be based on the general features of user requirements. These include suitable performance for the intended application, capability for incremental growth, reliability, and cost effectiveness. In general, these are competing features. High performance is costly, but modest performance with little expandability might be more expensive over time.

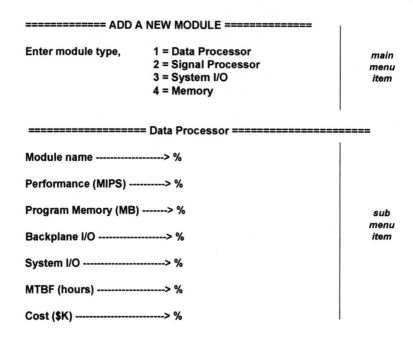

============ ADD A NEW MODULE =============

Enter module type,   1 = Data Processor   *main*
                        2 = Signal Processor   *menu*
                        3 = System I/O   *item*
                        4 = Memory

================== Data Processor =====================

Module name -----------------> %

Performance (MIPS) ----------> %

Program Memory (MB) -------> %   *sub*
                                                       menu
Backplane I/O -------------------> %   *item*

System I/O ----------------------> %

MTBF (hours) -------------------> %

Cost ($K) ------------------------> %

Figure 9-3   Sample menus for module entry.

The rules in this topic are similar to considerations that a system designer might perform manually. Since practical experience often is the basis for prioritizing these factors, a fuzzy logic, rule based system might be an efficient design engine. A later chapter revisits this idea.

9.1.2.1   *Performance*

The input performance requirements normally specify a function and the reserves that the system must meet. The reserves involve spare capacity to assure room for growth or changes in requirements during the development. Although some designers prefer to apply spare requirements at the computer system level, a more aggressive approach is to apply such requirements to the software processes assigned to each hardware element. This is the method used in this topic.

There are two very general categories of problems dealing with performance and architecture definition. The first class deals with new designs while the second class involves the reengineering of an existing design. The object of the first problem type is to map the software applications onto computing nodes and inventory the requisite hardware resources. The second issue might occur in conjunction with a major upgrade to the software of an existing network. The old software may have grown distorted over time with respect to resource needs. A remapping can balance these requirements to extend the life of an older architecture. Even though both problem domains are important, this discussion focuses on new designs.

The starting point for both problem classes is the same. This involves a capture of performance needs in terms of I/O, processor throughput, and

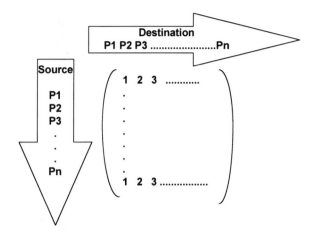

Figure 9-4 Definition of elements for matrix representation.

memory capacity. I/O is inherently different than the other two factors because it involves two system functions while throughput and memory size requirements are definable for each software process. Furthermore, for any pair of nodes engaged in I/O, the traffic might be bidirectional, initially with the first node the source, then with the first node the destination. For the sake of algorithm formulation, a matrix representation captures the I/O requirements cleanly.

Construct an $N \times N$ matrix, where N is the number of software processes (functions) in the network that must communicate. The net transfer requirements for this user defined message mix is:

$$L = \Sigma B_i R_{i..} \tag{9.1}$$

In this expression, the B matrix contains the bytes transferred from source to destination at a specified rate, as shown in Figure 9-4. The scalar Ri is the update rate for each matrix in Hz. The load matrix L is the sum of the products of rates and byte count matrices summed over all the possible rates. The transfer matrix is defined as:

$$T = L + L^T \tag{9.2}$$

The T matrix is symmetrical and represents the bidirectional transfer between any given two software processes in units of bytes per second. Once the I/O requirements are formalized, the communication between software processes can be optimized by the assignment of processes to nodes. For simplicity, the subsequent discussion assumes a *homogeneous system* that uses identical processor modules. This allows the analyst to assume that each node provides identical computing performance.

Several different experience derived methods can be the basis for node assignment. One example is hardware isolation to assure information secu-

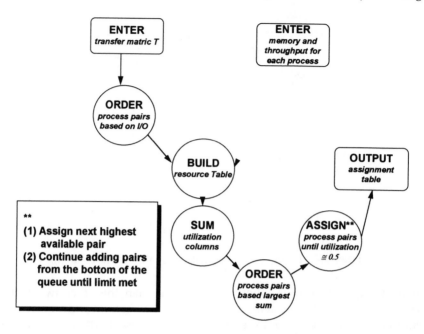

Figure 9-5   Node assignment algorithm.

rity. The classification level of the software process might be the basis of node assignments. Another factor could be fault tolerance, for which channel failure scenarios and software process criticality might dictate assignments. In the realm of performance analysis, two considerations represent the fundamental rules for node assignment.

The first heuristic is that software processes that communicate heavily should be assigned to the same node to minimize channel traffic. Sometimes, this is impossible because such processes might exceed the memory capacity or processor throughput of the node. This problem relates to the issue of *system constraints* and constitutes a refinement of the top level rules. This illustrates the more general principle of *requirements coupling* among the three aspects of processor throughput, I/O throughput, and memory size. Some algorithms function quite well when I/O loading is the predominant factor, but many practical engineering problems involve complex couplings.[2]

The second rule is to load each node up to some redefined utilization limit, typically 50% for new designs. This limit is derived from the reserve requirements that were mentioned earlier. Practical experience teaches that this approach can minimize the number of processing modules in the design, which helps to control cost.

Figure 9-5 shows the assignment algorithm based on these principles. The analytical calculations form the T matrix from the I/O inputs as discussed above. The process pairs are elements of a queue in order of utilization size, with the highest I/O utilization at the top. The I/O, throughput, and memory utilization are each summed and divided by the limiting utilization, which is usually 0.5. The highest of the three ratios gives the number

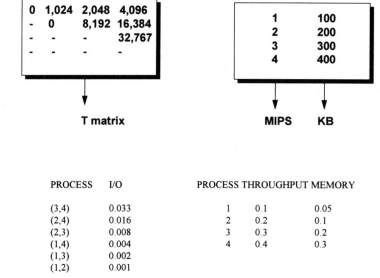

PROCESS	I/O		PROCESS	THROUGHPUT	MEMORY
(3,4)	0.033		1	0.1	0.05
(2,4)	0.016		2	0.2	0.1
(2,3)	0.008		3	0.3	0.2
(1,4)	0.004		4	0.4	0.3
(1,3)	0.002				
(1,2)	0.001				
SUMS:	0.064			1.0	0.65

Note: Assumes hardware capability of 1 MB/second, 10 MIPS, and 2 MB.

Figure 9-6   Algorithm example.

of nodes. Node assignment is based on whichever of the three factors is most critical as determined by the largest sum.

Figure 9-6 provides an example. The sum for throughput is the largest, so assignment is based on these values. Processes 1 and 4 are assigned to node 1, which brings the throughput estimate to the limit of 0.5. The external communication is 0.052 utilization while the memory utilization of node 1 is 0.35. The throughput utilization of node 2 is also 0.5; its external communication utilization is 0.052 (consistent since there are only two nodes), and the memory utilization is 0.3.

This example illustrates one quirk that may deserve special attention in implementations. Certain utilizations may represent "spikes" that are much higher than the other values, such as the first I/O entry in Figure 9-6. Although the algorithm functions properly, there is probably some practical extremum far below the required utilization limit that represents a reasonableness bound. For example, if all of the I/O values are increased by an order of magnitude, the I/O sum becomes 0.64, which still predicts the need for only two modules. The assignment based on the algorithm is the same as just discussed, but the I/O utilization is now 0.52, higher than the limit of 0.5. There is no obvious way to escape this problem since to assign the highest I/O users (3 and 4) to a node produced a processor throughput utilization of 0.7. The second pair on the I/O list (2 and 4) is no better, as it produces a processor throughput of 0.6. The third pair is (2,3) which produces the same answer as before.

The solution to such problems is a combination of validity checking of inputs and statistics. Above some arbitrary utilization, perhaps 0.25, one could calculate both the mean and standard deviation of the values. If values lie more than two standard deviations from the mean, there may be a problem with the inputs.

Another approach is validation against a model. Return to the issue of coupling among requirements. A performance characterization of the intended application must draw from some underlying theoretical model. A simple but powerful model is to assume a linear relationship:*

$$T = T_0 \left(1 - I/I_0\right), \tag{9.3}$$

where T = processor throughput (variable)
T_0 = processor throughput (maximum attainable)
I = I/O bandwidth (variable)
I_0 = I/O bandwidth (maximum attainable).

The utilizations are T/T_0 and I/I_0. Given a requirement that $I/I_0 = 0.33$ for the process pair (3,4), the maximum achievable throughput utilization in this model is 0.66. However, the processor throughput input is 0.7 for these two processes which indicates, in essence, that the required throughput is unachievable. Clearly, the inputs are invalid!

One final issue is I/O bandwidth. Notice in the previous discussions of I/O utilization that the reserve capacity requirement forces each I/O utilization value to be less than 0.5. A naive bandwidth estimate might be the sum of all the values for a node, which could be higher than this. However, a calculation of channel capacity requires not just the I/O utilizations, but also the message scheduling that is implicit in the theory. That is, each pair of communicating nodes controls the channel for some length of time and then relinquishes it to another pair. The performance factor is *end to end message latency*, which defines how much time evolves from the request for a transmission to its arrival at the destination. The bandwidth consumption varies with time depending on which pair of processes is communicating.

9.1.2.2 *Incremental growth*

Growth capability is another important characteristic of the design. Performance, architectural topology, and network communication operation are important factors that influence the capability of a design to grow.

Performance impacts due to an incremental upgrade may be the easiest of the factors to study. The node assignment analysis of the last discussion applies directly. The analyst can add the new performance requirements and simply remap the software processes to determine the number of nodes needed. This might be performed with an entirely new system definition derived from all of the software processes.

Alternatively, the growth analysis could be iterated from the baseline. The analyst might attempt to measure growth capability by simply duplicating the existing system functions instead of extracting a point design for

---

* Despite the simplicity of this formulation, it has a theoretical basis. The author has tested this model on numerous computers with excellent results.

every new functional input. In this method, the analyst might add a duplicate of every fifth process and remap, then every fourth, third, and so on until all processes are duplicated. The number of nodes needed for each implementation can then be plotted against performance requirements and cost. This gives a direct measure of growth in terms of conventional metrics such as $/ MIPS. It also provides insight into the limit points beyond which new computers, not simply new modules, are required.

The architectural topology is another consideration. Although topological definitions are outside the scope of the current discussion, some simple examples show the importance of this factor. Consider a linear topology in which each node has only two nearest neighbors, one on each side. At some point, physical and protocol limits prevent the addition of more nodes and a gateway to another network is needed. Obviously, the gateway hardware and software, as well as the new network configuration, add considerably to the system cost. Therefore, these limiting points may represent the practical growth limits of the architecture.

The network communication method relates to system complexity and message transmission. Although a description of the model types is also beyond the scope of this discussion, the essence of the issue is whether messages are *blocking* or not blocking. Message blocking occurs when the data processor must wait for a hardware communication resource to become free before it can continue operation. The classic example is a multiple ported, shared memory architecture with write locks. The blocking behavior is a characteristic of both the messaging protocol and the hardware implementation. Performance penalties due to blocking can be considerable for large numbers of nodes attempting to access a shared resource.[3]

Some real-time embedded system designs must remain operational for many years. In such cases, it may not be cost effective to replace an entire computer system. This cost often stems from the expense of managing many different configurations of hardware in terms of spare parts, personnel and training, software configurations, and other factors. In such situations, incremental growth may be an important consideration.

9.1.2.3 Hardware reliability

A system specification might characterize the hardware reliability requirement in terms of hours between failures. Simple probabilistic models allow calculation of this result once the hardware module types, counts, and groupings are available from the performance analysis.

Reliability simply means the frequency that the system needs repairs. The probabilistic reliability model provides two outputs. The first is the mean time between failures (MTBF) while the second is the mean time to repair (MTTR). A specific hardware module type has unique values for these parameters. The system MTBF is calculable from the MTBF of individual modules:

$$1/\lambda = \Sigma 1/\lambda_i, \quad i = \text{module numbers}, \tag{9.4}$$

in which λ is the failure rate, or inverse of the MTBF.

This system MTBF model assumes that each module must function in order that the system operate. If a designer uses hardware redundancy to lower the risk of failure for critical functions, one simply multiplies a single module MTBF with the number of occurrences for that function.

System availability is a related issue. One can model this as simply the ratio of MTBF to the sum of MTBF and MTTR:

$$\text{availability} = \text{MTBF} / (\text{MTBF} + \text{MTTR}). \quad\quad (9.5)$$

This can be expressed as a percentage that indicates over the life of the host system, how many times the computer must be replaced.

The reliability model also relates to life cycle cost impacts. The original cost of the computer unit might be small in comparison to the repair bills for an unreliable design. At some point it becomes more cost effective to simply replace the unit than repair it.

9.1.2.4   Power, weight, volume

There may be additional input requirements related to physical aspects of the design. These might include the power dissipated by the computer, power consumed by the computer, total weight, and volumetric size. These factors usually must fall below some limit in order to fall within the constraints of the host system.

The power dissipated usually is expressed in watts. This is a measure of how much heat the computer can produce. This factor might be important so that the host system designer can avoid specialized cooling methods such as rapid flow rates of cool air or liquid cooling. The actual power consumed depends on the hardware scenario. Consequently, the requirement may be stated in terms of worst case conditions. This might involve holding all signal lines at high voltage and assume maximum current draw. Often, it is impossible to have all inputs high simultaneously so this scenario might need tailoring. The purpose of this requirement is to assure that the electrical power source of the host system can provide sufficient current and voltage draws.

The total weight is simply the sum of all the hardware modules, cabling, and enclosure. The purpose of this requirement may derive from a maintenance approach or physical constraints of the host system.

Finally, requirements also limit the size of the computer. In addition to overall volume, each dimension usually must fall below some maximum limit to assure that the design is usable in the host system.

As with reliability, these factors are easily calculable once a design baseline emerges from the performance analysis.

9.1.2.5   Rule based design system

Figure 9-7 shows a rule based system that formulates a design baseline. The values in the right hand gray box are input requirements for the area shown adjacent in the left hand box. The analysis section calculates values for each factor by which to evaluate the inequalities. For example, if the performance reserve requirement is 50%, analysis similar to that of Section 9.1.2.1 might

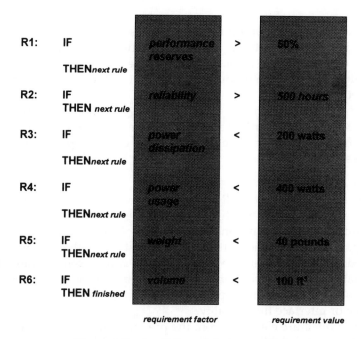

Figure 9-7 A rule based design method.

produce an output. The next rule could be evaluated using the methods of Section 9.1.2.3. In a similar manner, there is an analysis package for the other factors whose outputs are the manner by which one evaluates the inequalities. In this simple model, any if-then evaluation that fails invalidates the design entry.

This system is an expert system because the data values are decoupled from the decision logic. A more conventional algorithm might have these values hard coded into the code or read in from a data file.

There may be other factors as well. System cost might be evaluated and candidates ranked to minimize this factor. The definition of incremental growth might be formalized and this factor added. As is true with all good design efforts, though, the baseline must be derived from customer requirements instead of arbitrary decisions of the design team.

A similar system could be formulated to make recommendations about office desktop computers. The system might ask the user about the intended application (the software that will run on the system). A database of requirements for various types of applications could be available to the analysis code. This database might specify the memory size, hard disk capacity, graphics capability, and similar elements. Prices for systems and components might also be available to the analysis section to provide the customer with a cost quote.

In order to perform trade studies, fuzzy logic may be more appropriate to extract the best design. A membership function might be simply a "triangle function" that peaks at the target value. For example, the output of the membership function for the performance parameter might produce a value of 0 for 0%, 1 for 50%, and 0 for 100%. Line segments join these points to produce intermediate outputs. Therefore, the output from each rule is a value between 0 and 1. The designer might formulate a TPM (see Chapter 1) from the product of each such output. The next topic addresses this approach in more detail.

This portion of the chapter provides some evaluation methods and rules for decision making regarding a design baseline. The next topic discusses the outputs from the design engine. The remainder of the chapter provides some examples.

9.1.2.6   VP system

It is possible to construct a VP once the design engine formulates a level 0 architecture block diagram. This approach relies on behavioral simulations of the architecture elements and automated tools that produce engineering drawings, detailed specifications, component lists, and other design information needed to manufacture the system. The VP is also useful in trade studies to validate both the design and the input requirements. This discussion examines such applications, especially those involving trade studies for performance or interface functionality.

The architecture elements from the previous stage are the various bus types and module definitions. The issue in establishing a VP involves formulating timing relationships between behavioral simulations for such interfaces. The bus protocol behavioral simulator contains the various system states and timings for a particular interface. However, experience teaches that accurate performance estimates involve much more than the time needed to gain control of the bus and transmit the message from source to destination at the module interfaces. Instead, each module adds a delay as it transfers the information from its I/O interface to the memory system in which the application software executes. The module might process this information and then retransmit it to another module. Eventually, some module transmits the information to an I/O module. The I/O module transfers the information internally and then transmits the data to an external source.

The descriptions of the delays involve several factors. The first is the *burst rate*, which describes how quickly a particular module can transmit a message to the destination. In principle, this value is a function of the message size and the physical distance to the destination. Ordinarily, the message scheduling delays overwhelm these factors. The message scheduling delays are a function of the messaging execution environment, since a module may be forced to wait to transmit while another module completes its transaction. Finally, the software scheduling delay involves the rate at which the application program updates the information. Usually one does not schedule messages faster than this software update rate, although there are exceptions.

Therefore, the *node delay* for an interface depends on:

- Software scheduling rate (information update rate in Hz)

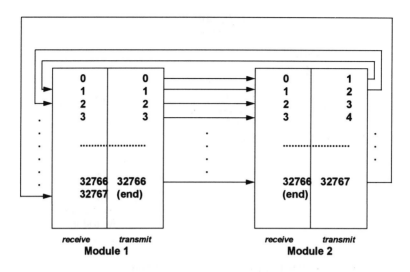

Figure 9-8 Message latency example.

- Host side burst rate (time needed to move the information from application memory to I/O memory and trigger the I/O controller)
- Any delays added by processing the information at the node.

Consider a simple example. Two data processor modules perform a loop back test using the parallel data channel. The first module schedules a single message that contains one data word. Initially, this data word is zero. The first module transmits this message at 20 Hz. The second module receives the data word and simply adds one to its value and retransmits as quickly as possible. That is, the second module polls its interface awaiting the arrival of a new data package. The first module compares the received message to both the message it just transmitted and the message it last received. The test fails whenever the received value at the first module equals either of these two values (they should be equal). The data word is the loop counter and measures the number of successful transmissions. At some point, the test must terminate so that this counter does not overflow. This value usually will be 32,767.

Figure 9-8 shows the operation of this loop test. The boxes indicate the hardware modules. Each module has both a receive buffer and a transmit buffer. The host side delays involve the transfer time between these buffers. These delays include any time to service the I/O interrupt and to perform the copy operation. In addition, the first module requires a 20 Hz (50 ms) delay between these two buffers because of software scheduling. Finally, the second module includes a processing delay because it updates the data value. The lines between the modules indicate transfers on the parallel bus system.

This example illustrates a pitfall of many simulation efforts. Some analysts provide latency predictions on the basis that each module is ready and waiting for access to the bus. This means that there is no latency modeled

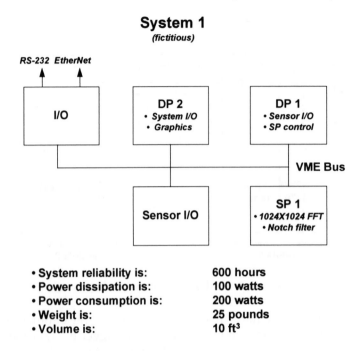

Figure 9-9   Sample design output.

between the receive buffer and the transmit buffer. Sometimes this is a good approximation, but usually it is not. Unfortunately, this approximation also modifies the message scenario in the bus simulator to make the results even worse. A realistic latency check must account for host side delays.

The analysts must provide the software scheduling delays, but the hardware delays can be simulated. The design of a new hardware module usually involves behavioral simulation early in the process. Although such models are often too complex to couple directly to bus simulations, it is possible to build timelines for a small number of events. In this case, the system integrators require the latency between the I/O interface and the application program.

Suppose that a group of analysts wish to use the loop test as a system performance test. The measurement involves the elapsed time between the start of the test and the output of the loop counter. For a block diagram such as Figure 9-9 (next topic), the loop test executes over the parallel bus (VME) and the terminal output uses the serial bus (RS-232). A VP of this system might involve the receive and transmit buffer delays for each bus type just discussed. The analyst only needs to print out the simulation elapsed time after completion of the test.

This discussion describes how the outputs of the design process might be used in trade studies. Other on-line tools might produce various data package elements for the design. The result of this automated approach is cost savings in both the trade studies and the data package production. Many of these tasks are now performed manually at great expense. Chapter 10 continues this discussion.

9.1.3   Configuration outputs

The design system output contains a list of the module types, counts, and interconnects for each candidate. Figure 9-9 shows an example of how this information might be represented graphically. The figure shows a *level 0 architecture block diagram* that shows all the functional module types and interfaces. The power supply modules might not appear on such diagrams as they usually do not require digital interconnects. However, these modules must be included in the estimates.

The design engine might produce several candidates that meet the requirements. The analysis could produce a rating of the candidates based on the top level requirements described in the last discussion. This might involve a postprocessor that constructs lists that the various behavioral simulations might need for initialization. This application might be suitable for fuzzy logic since the rules for establishing priorities ought to be clear, but the precise boundaries may not be evident. The next example illustrates these ideas.

9.2   Fuzzy rule based design

The last topic addressed the interfaces and functions of a system design engine. This discussion extends the methods to include fuzzy rule based design. The bulk of the discussion focuses on a simple example, but also includes a summary of the general case.

Consider a homogeneous, parallel processing example. The system design involves a collection of data processors that must each support a concurrent software application. The architecture selection criteria are system performance, reliability, and cost. An exponential performance model provides a simple estimate of available performance. In this model, each new module adds 80% of the performance of a single module to the system performance. The MTBF assumes that each module must function for the application to execute normally. Therefore, the system MTBF is simply the single module MTBF divided by the module count. Finally, the system cost is the single module cost multiplied by the module count.

The characteristics of a single data processor might be:

Performance: 5 MIPS
MTBF: 15,000 hours
Cost: $15,000.

The system requirements might be:

Performance: 15 MIPS
MTBF: 3,000 hours
Cost: $50,000.

The membership function for performance might be a sawtooth pattern:

CP = calculated performance of the candidate system configuration,
RSP = required system performance

$$\mu_{performance} = \begin{cases} CP/RSP, & \forall 0 \le CP \le RSP \\ 2 - CP/RSP, & \forall RSP \le CP \le 2*RSP \\ 0, & \forall CP > 2*RSP. \end{cases}$$

A similar membership function applies to system cost if the designer must meet a cost target. One can generalize this assumption for use in the situation in which the designer must not exceed a cost threshold. The next element of this discussion presents this generalization. However, the membership function for MTBF might be pegged to 1.0 whenever calculated reliability exceeds the requirement. This "up ramp" membership function might be more appropriate for a high reliability system. Notice that all of the output values of the membership functions lie in the range [0,1]. The designer might formulate a TPM as the product of the values of the membership functions. The closer to 1 that the TPM approaches, the better the design solution.

Table 9-1 shows the impact of varying each parameter. The entries of the table are values of the TPM discussed above. One can formulate estimates of the module counts from the baseline system requirements and the algorithms cited above for reliability, cost, and performance:

MTBF > 3,000 $\Rightarrow$ # of modules = [1,..,5]
Cost < $50,000 $\Rightarrow$ # of modules = [1,..,3]
Performance > 15 $\Rightarrow$ # of modules $\ge$ 4.

This is consistent with the number of modules that the TPM selects. However, the TPM offers much more information than the optimal module configuration. Notice that as the system requirements increase, the value of the TPM decreases. This indicates that the requirements are not well specified. Below some threshold (perhaps 0.5) the fuzzy rule based design system might warn of inappropriate requirements.

This process involves manual estimation of the optimal TPM, but an automated procedure is possible. The value of the TPM can be a control input to an expert control system. The controller varies (in this case) the number of modules to produce an optimal solution for a given set of system requirements. The expert system could then vary the requirements to verify that they are well formulated. If they are not, the expert system could recommend new system requirements. Alternatively, one could view TPM < 0.5 as the need to select a new module type.

Many different module designs might be available to the system integrator. The TPM can select the optimal module configuration in the manner just discussed. Table 9-2 shows the impact of varying the module characteristics. This example assumes a linear relationship between performance and module cost. The TPM selects the obvious choice which is a single module that

Table 9-1 Impact of Varying Performance System Requirements

Module cost = $15,000; Module MTBF = 15,000; Module performance = 5 MIPS

Case 1: System cost < $50,000; System MTBF > 3,000 hours; System performance variable

Number of modules	Reliability[a]	Cost[b]	15 MIPS	30 MIPS	45 MIPS	60 MIPS
1	15.000	15	0.10	0.05	0.03	0.03
2	7.500	30	0.36	0.18	0.12	0.09
3	5.000	45	*0.78*	0.39	0.26	0.20
4	3.750	60	0.69	*0.45*	*0.30*	*0.23*
5	3.000	75	0.30	0.35	0.23	0.18
6	2.500	90	0.06	0.14	0.09	0.07
7	2.142	105	0.00	0.00	0.00	0.00
8	1.875	120	0.00	0.00	0.00	0.00
9	1.667	135	0.00	0.00	0.00	0.00
10	1.500	150	0.0	0.00	0.00	0.00

Case 2: System cost < $50,000; System MTBF variable; System performance = 15 MIPS

Number of modules	MIPS	Cost	MTBF = 3,000	MTBF = 6,000	MTBF = 12,000	MTBF = 18,000
1	5	15	0.10	0.10	0.10	0.10
2	9	30	0.36	0.36	0.30	0.23
3	13	45	*0.78*	*0.65*	*0.43*	*0.33*
4	17	60	0.69	0.43	0.29	0.22
5	21	75	0.30	0.15	0.10	0.08
6	25	90	0.06	0.03	0.02	0/01
7	29	105	0.00	0.00	0.00	0.00
8	33	120	0.00	0.00	0.00	0.00
9	37	135	0.00	0.00	0.00	0.00
10	41	150	0.00	0.00	0.00	0.00

[a]Reliability is given in thousands of hours.
[b]Cost is given in thousands of dollars.

can deliver exactly the system performance and exceeds the system reliability requirements. Additional performance beyond this simply increases the module cost which the TPM values demonstrate.

Minimum cost could be a consideration instead of a cost target. For this case, the membership function might be a ramp function:

CC = calculated cost
RSC = required system cost threshold

$$\mu_{cost} = \begin{cases} 1 - CC / (2 * RSC), & \forall \, 0 \leq CC \leq 2 * RSC \\ 0, & \forall \, CC > 2 * RSC. \end{cases} \quad (9.6)$$

Table 9-1 (continued)   Impact of Varying Performance System Requirements

Case 3: System cost variable; System MTBF > 3,000 hours; System performance = 15 MIPS

Number of modules	MIPS	Cost = $50,000	Cost = $100,000	Cost = $150,000	Cost = $200,000
1	5	0.10	0.05	0.03	0.03
2	9	0.36	0.18	0.12	0.09
3	13	*0.78*	0.39	0.26	0.20
4	17	0.69	*0.52*	*0.35*	*0.26*
5	21	0.30	0.45	0.30	0.23
6	25	0.06	0.25	0.17	0.13
7	29	0.00	0.05	0.03	0.03
8	33	0.00	0.00	0.00	0.00
9	37	0.00	0.00	0.00	0.00
10	41	0.00	0.00	0.00	0.00

Case 4: System cost variable; System MTBF variable; System performance = 15 MIPS

Number of modules	MIPS	Cost = $50,000 MTBF = 3,000	Cost = $100,000 MTBF = 6,000	Cost = $150,000 MTBF = 9,000	Cost = $200,000 MTBF = 12,000
1	5	0.10	0.05	0.03	0.03
2	9	0.36	0.18	0.10	0.06
3	13	*0.78*	*0.33*	*0.14*	*0.08*
4	17	0.69	0.33	0.14	0.08
5	21	0.30	0.23	0.10	0.05
6	25	0.06	0.13	0.06	0.03
7	29	0.00	0.02	0.01	0.01
8	33	0.00	0.00	0.00	0.00
9	37	0.00	0.00	0.00	0.00
10	41	0.00	0.00	0.00	0.00

[a]Reliability is given in thousands of hours.

[b]Cost is given in thousands of dollars.

Table 9-3 shows the TPM values as the system performance requirement varies with the membership function above. The new cost membership function treats any module count below four equally, which allows performance to drive the count selection. Comparison of the results of Tables 9-1, 9-2, and 9-3 bears out this trend.

The example is simple enough to analyze in a closed form algorithm for the TPM. An algorithm might be the basis of a more classical approach such as a state vector method. The values of the three membership functions for the hyperspace area nearest the origin are:

Table 9-2 Evaluation of Module Candidates

Number of modules	1 MIPS	5 MIPS	10 MIPS	15 MIPS	20 MIPS	25 MIPS	30 MIPS
1	0.01	0.10	0.40	0.90	0.53	0.17	0.00
2	0.03	0.36	0.64	0.04	0.00	0.00	0.00
3	0.50	0.78	0.05	0.00	0.00	0.00	0.00
4	0.80	0.69	0.00	0.00	0.00	0.00	0.00
5	0.10	0.30	0.00	0.00	0.00	0.00	0.00
6	0.11	0.06	0.00	0.00	0.00	0.00	0.00
7	0.11	0.00	0.00	0.00	0.00	0.00	0.00
8	0.13	0.00	0.00	0.00	0.00	0.00	0.00
9	0.15	0.00	0.00	0.00	0.00	0.00	0.00
10	0.16	0.00	0.00	0.00	0.00	0.00	0.00

Table 9-3 TPM Results for Minimum Cost Criteria

Module cost = $15,000   System cost < $50,000
Module MTBF = 15,000   System MTBF > 3,000 hours
Module performance = 5 MIPS   System performance variable

Number of modules	Reliability[a]	Cost[b]	15 MIPS	30 MIPS	45 MIPS	60 MIPS
1	15.000	15	0.28	0.14	0.09	0.07
2	7.500	30	0.42	0.21	0.14	0.11
3	5.000	45	*0.48*	*0.24*	*0.16*	*0.12*
4	3.750	60	0.35	0.23	0.15	0.11
5	3.000	75	0.15	0.18	0.12	0.09
6	2.500	90	0.03	0.07	0.05	0.03
7	2.142	105	0.00	0.00	0.00	0.00
8	1.875	120	0.00	0.00	0.00	0.00
9	1.667	135	0.00	0.00	0.00	0.00
10	1.500	150	0.0	0.00	0.00	0.00

[a]Reliability is given in thousands of hours.
[b]Cost is given in thousands of dollars.

$$\mu = \begin{cases} x1/S1 & \forall 0 \le x\,1 \le S1 \\ 1 & \forall 0 \le x\,2 \le S2 \\ x3/S3 & \forall 0 \le x\,3 \le S3 \end{cases} \tag{9.7}$$

These expressions represent the membership functions that the introduction of this topic describes:

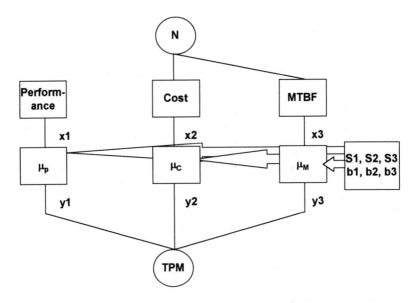

Figure 9-10   Block diagram of the TPM calculation.

$$x_1 = \pi\left[\left(1 - r^N\right)/(1 - r)\right] \qquad \text{(performance)}$$

$$x_2 = \rho N \qquad \text{(system cost)} \qquad (9.8)$$

$$x_3 = \lambda/N \qquad \text{(MTBF)}$$

in which N is the number of modules and (π,ρ,μ) are the single module performance, cost, and MTBF values. Therefore, the TPM is:

$$TPM = \mu(x1) * 1 * \mu(x3)$$

$$= \left[\pi\lambda/(S1\,S3)\right]\left[\left(1 - r^N\right)(1 - r)/N, 9\right] \qquad (9.9)$$

which produces an exponential decay of the TPM with increasing module count. There are nine different combinations of algorithms based on the condition of the logic test $0 < xi < Si$, $Si < xi < 2Si$, or $xi > 2Si$ for $i = 1,2,3$.

Figure 9-10 shows the logic tree for these calculations. The module count (N) is the data input. The inference engine is forward chained (data driven) through the system requirements functions. If this system were implemented as a neural network, the node weights would be S1, S2, S3.

Another issue is how to iterate the design using some type of feedback. In this case, the input is the module count. The initial estimate that such methods require could be based on simply dividing the system performance requirement by the module performance provided. The design engine must calculate TPM values on each side of this initial estimate to establish a trend. The maximum TPM value marks the recommended configuration.

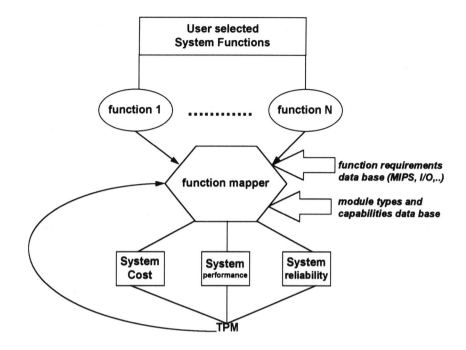

Figure 9-11   Rule based design engine.

A complete system design engine might resemble the structure of Figure 9-11. The user input is a set of top level functions that the system must support. In avionics jargon, these are *configuration items*. Historical data provide estimates for the data processing, signal processing, I/O, and storage size requirements. Another data structure might contain module capabilities for particular vendor designs with respect to these areas of estimation. The estimates, input function list, and module types are the basis of a node assignment engine similar to the example discussed in Section 9.1. The module counts, individual costs, and module reliabilities could feed a rule based system similar to that of Figure 9-1. This system produces the TPM which the rule based system can iterate as just discussed.

In order to complete a realistic design engine, one needs to add the remaining rules for memory size and I/O performance. The membership functions are identical to the performance function discussed previously. However, although this example is intentionally simple, it does illustrate use of membership functions and design iteration using feedback.

9.3   New design case study

This example provides more details of the design analysis process. It also provides some example outputs for some simple input requirements. The software process to node assignment algorithm may be the most complicated portion of the analysis. Therefore, this discussion focuses on a more complex message mix.

destination⇒ process	1	2	3	4	5	6	7	8	9	10	11	12	13	14	15	16

source process
⇓

	1	2	3	4	5	6	7	8	9	10	11	12	13	14	15	16
1	0	15	20	25	30	25	20	15	10	5	10	15	20	25	30	25
2		0	20	15	10	5	10	15	20	25	30	25	20	15	10	15
3			0	20	25	30	25	20	15	10	5	10	15	20	25	30
4				0	25	20	15	10	5	10	15	20	25	30	25	20
5					0	15	10	5	10	15	20	25	30	25	20	15
6						0	10	5	10	15	20	25	30	25	20	15
7							0	10	5	10	15	20	25	30	25	20
8								0	15	10	5	10	15	20	25	30
9									0	25	20	15	10	5	10	15
10										0	15	20	25	30	25	20
11											0	15	10	5	10	15
12												0	20	25	30	25
13													0	20	25	30
14														0	25	20
15															0	15
16																0

Notes:

(1) The matrix is symmetric, redundant entries are omitted for clarity.
(2) The entries are percentages instead of utilizations (i.e., divide by 100).

node⇒	1	2	3	4	5	6	7	8	9	10	11	12	13	14	15	16
processor utilization⇒	10	15	20	25	30	25	20	15	10	5	10	15	20	25	30	25
memory utilization⇒	10	15	20	25	30	25	20	15	10	5	10	15	20	25	30	25

Figure 9-12   New design example transfer matrix.

Consider the design of a new computer network. The first step for an analyst in establishing utilizations is formulation of typical and worst case software process requirements. Suppose, for example, that each software process must transfer from 1 to 10 messages whose maximum length is 32 words. The applications that generate this I/O must perform various user initiated operations to generate these messages. Figure 9-12 shows the utilizations for this example.

As the number of nodes becomes larger, the utilization information becomes more difficult to present tractably. Notice that for N nodes, the number of I/O utilization entries is $N(N-1)/2$, from the combination of N items in two ways. For this example, $N = 16$, so the number of process pairs is 120. This is clearly large enough to warrant some type of automated analysis.

WARNING! invalid MIPS for 4 and 3
utilization is 32
MIPS requirements is 70

PROCESS	THROUGHPUT	MEMORY
1	10	5
2	20	10
3	30	20
4	40	30

PROCESS	I/O
(4 , 3)	32
(4 , 2)	16
(3 , 2)	8
(4 , 1)	4
(3 , 1)	2
(2 , 1)	1

SUMS	63

PROCESS	NODE
1	0
2	1
3	1
4	0

Figure 9-13 Output for the simple example.

As an introduction to the automated method, reconsider the data of Figure 9-6. For comparison, Figure 9-13 shows the output of the automated system for these data. The utilizations are percentages, not fractions. The pattern of this output is similar to the analysis presented in the last section. Notice also the warning message that is generated from a check using Equation 9.3. The output format of the more complex example of Figure 9-9 is similar.

Figure 9-13 shows the output from the analysis section of the expert system. As the message indicates, the I/O loading is the assignment criterion. There are five nodes for the 16 processes, and assignments are formulated to meet the 50% reserve requirement for I/O. However, the processor throughput and memory utilizations for the five nodes are (60, 65, 65, 70, 40), so some additional assignment resolution is necessary.

The assignment algorithm is easily refinable in terms of the other utilizations. Although minimizing I/O utilization is the first criterion, the processor and memory utilization must also be less than 50%. Figures 9-14 and 9-15 display the two cases for the inputs of Figure 9-12. Without the additional utilization requirements of throughput and memory, only five nodes are required, but seven are required with the additional restrictions. Figure 9-13 presents the node assignments for each case and Figure 9-15 shows the

Case 1:

PROCESS	NODE
1	1
2	1
3	2
4	3
5	1
6	2
7	4
8	5
9	2
10	1
11	2
12	3
13	4
14	3
15	4
16	5

Case 2:

PROCESS	NODE
1	1
2	1
3	2
4	2
5	3
6	4
7	5
8	5
9	5
10	5
11	6
12	3
13	4
14	6
15	7
16	7

Figure 9-14   Node assignment in the second example.

utilizations. Notice in case 1 that there is excessive utilization. To summarize, case 1 is based solely on trying to assign processes by meeting a 50% I/O loading, while case 2 is based on minimizing I/O usage while preserving the 50% reserve in processor and memory utilization.

One interesting aspect of this approach is its capability to abstractly handle either networks of subsystems or hardware modules. Figure 9-16 shows an example of the latter. In this case, a navigation processor accepts sensor inputs from both self-contained, inertial sensors and radio navigation interfaces. The I/O utilization is a function of the bandwidth of the backplane bus, the message sizes, and update rates. For example, if a realistic bandwidth is 4 MB/s and 6 messages of 10 words each are transmitted at 80 Hz,

Case 1:

NODE	MIPS	KB
1	60	60
2	65	65
3	65	65
4	70	70
5	40	40

Case 2:

NODE	MIPS	KB
1	25	25
2	45	45
3	45	45
4	45	45
5	50	50
6	35	35
7	55	55

Figure 9-15   Node utilizations.

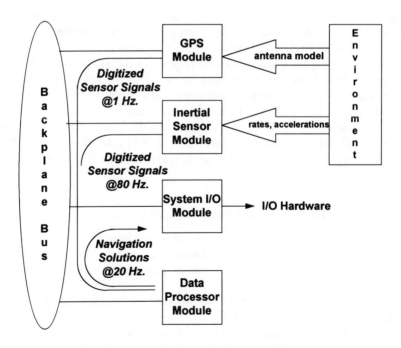

Figure 9-16   Navigation subsystem example.

the utilization for the inertial interface is 0.00024. The processor and memory throughputs may be more straightforward to estimate, but the reader is cautioned, as always, against simple models such as scaling based on MIPS ratings. The extensive floating point calculations dealing with the navigation digital filter are likely to make processor throughput the critical factor for this example.

An older implementation of this function might involve a federated architecture in which each separate computer couples with a single sensor interface for each subfunction. These computers might be connected with a high speed, serial bus. In such systems, each node is an entire computer, not simply a hardware module. Formulation of the utilizations, though, is similar to the process just described. Once the utilizations are available to the analyst, the node assignment algorithm operates as before.

There are other factors that may impose additional partitioning requirements, as mentioned in the previous section. For example, a fault tolerance requirement may dictate a hot or pooled spares approach for the hardware. In the hot spares method, the system hardware duplicates (2*N redundancy) the hardware and software of any critical functions. The pooled spares approach is available in homogeneous systems. In this method, the system provides a single additional node (N + 1 redundancy) onto which software of any failed node can be loaded and started. The pooled spares approach requires fewer resources, but the start-up times in the event of a failure may be prohibitive.

Another factor is information security. In older systems, the security methods often assumed hardware isolation of key functions. Newer systems can obviate this limitation, though, with new operating system approaches and the speed of high performance RISC processors.

A final concept is distributed, heterogeneous networks in which the data processors that drive the nodes may be of different types. The theory and examples until now assume a homogeneous system. Usually, for new designs of embedded systems, homogeneous architectures can be more cost effective due to the economy of scale derivable from a common modules approach. However, existing systems may be formulated of many different computer types. Enhancements to such systems often focus on I/O, as the next example illustrates.

9.4   *Architecture reengineering case study*

Architecture reeingineering involves reallocating hardware resources to provide additional growth capacity. In the simplest conceptual terms, designers often sell reengineering to customers as primarily a software effort with minimal hardware impact.

A situation of this type might be much more familiar than first impressions might suggest. Imagine a small office that contains two work stations for logic design. Of course, each has a network interface installed to simplify file exchange. One of the two computers contains the master files for the design project. Two years later, the business has grown to need ten work stations. However, the same file management system is in place that forces users to submit baseline captures to a single computer on the network. This central computer quickly becomes a bottleneck and slows the entire network.

Perhaps the obvious solution to this problem might be a more powerful central computer. However, the customer might be concerned that an identical situation could occur 2 years later when 15 work stations are on line. At some point, limitations in the underlying network preclude any further growth. The proximity of this end stage is a critical issue for business managers as it impacts equipment purchase schedules, training budgets, and worker productivity. The customer needs an accurate answer from the architecture reengineering team to this key question of capacity.

Eventually, it may become obvious that the addition of more nodes is no longer possible. However, financial considerations may dictate that the existing computers not be discarded, but upgraded. Several approaches are possible at this point. First, small groups of work stations may be decomposed into network clusters with cluster controllers that communicate to the network central controller. Another approach might be a distributed computing approach that spreads pieces of the control and management operations among the many work stations.

The theory of Section 9.1.2 provides a point of departure for the analysis involved in reengineering. Suppose that the utilizations shown in Figure 9-17 represent the existing network system to modify. In this case, the goal is not to calculate the minimum number of processing nodes for a software architecture, but rather to eliminate the bottleneck with a software remap. The utilization (load) matrix shown at the bottom of Figure 9-17 may more strongly emphasize this communication bottleneck.

The nature of a central controller may mask the true routing of messages in this architecture. It is likely that a designer submits files from one node to the controller, which subsequently releases the same messages to another node. This has the effect of doubling the message traffic in comparison to the situation where no intermediate controller node is involved.

A distributed solution to this network can be much more efficient. Each node has a configuration management software system that tracks and archives the current design files. Users from anywhere in the office network can obtain a directory or request files from this management system coupled to the network directory software. In the old system, the central controller periodically builds a design capture by uploading all the current baseline files. In the distributed approach, it is no longer necessary for a node to submit a request to the central controller to download a file from its storage. This can decrease channel utilization substantially, as previously discussed. The trade-off is that each node has a higher processor utilization because it is running more software. Establishing a balance between processor and I/O utilization is never easy, but for older systems, this becomes the essence of reengineering.

This brings the discussion back to the three central elements of system utilization. These are processor, I/O, and memory usage. The three are not independent, though (see Equation 9.3), and this can be a valuable observa-

Utilization Table

PROCESSES	I/O		PROCESS	THROUGHPUT	MEMORY
(3,4)	0.3		1	0.1	0.3
(2,4)	0.3		2	0.1	0.3
(2,3)	0.05		3	0.1	0.3
(1,4)	0.3		4	0.4	0.4
(1,3)	0.05				
(1,2)	0.05				
	----			----	----
SUMS:	1.05			0.7	1.3

Note: Assumes hardware capability of 1 MB/second, 10 MIPS, and 2MB.

Transfer Matrix

$$
T = \begin{vmatrix}
0 & 0.05 & 0.05 & 0.3 \\
0.05 & 0 & 0.05 & 0.3 \\
0.05 & 0.05 & 0 & 0.3 \\
0.05 & 0.05 & 0.05 & 0
\end{vmatrix}
$$

Figure 9-17 Reengineering example.

tion in a reengineering effort. As just mentioned, the general strategy in reengineering is that processor utilization must be traded for I/O utilization. The relationship between these two factors may not be linear due to the architecture and operating system implementation. However, the two variables are not independent and the analysts must establish the relationship early in the effort to optimize the final design.

Reengineering is already an important discipline in embedded computing. Often, physical limitations such as the electrical interfaces, maintenance strategies, and business policy demand that the same computing elements be used in a product for a decade or more. Given the demands of software developers, this can produce the situation just described in approximately 5 to 7 years. This method is also important for commercial users, who are building increasingly larger networks to support business and engineering activities. Eventually, a "tipover" point occurs in which it is no longer financially possible to replace all of the hardware and software. An expert system for VP can be a valuable tool in evaluating the various options to solve this situation. In addition, an OOD design of the expert system could simplify transition to a new baseline when a hardware upgrade occurs. VP and OOD are key methods to assist the analyst at quickly deriving accurate solutions.

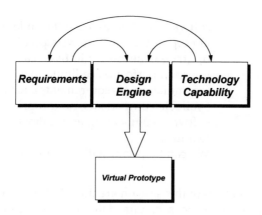

Figure 9-18 Elements of a VP system.

9.5   *Status of automated design*

This chapter presents algorithms that replicate the most significant, yet elementary, design rules for computer systems design. These rules serve well in the formulation of both new designs and reeingineered systems. This discussion addresses the computing requirements necessary to run such design rules, as well as the place of such a system in the overall design process.

The significance of the design rules is to serve as a model for the iterative process shown in Figure 9-18. This diagram summarizes some of the top level features of Figure 1 in the introduction to Part II. Each element of this diagram must function autonomously, yet integrate seamlessly with the other elements to form an automated design system. The VP output is the initial baseline that analysts can test in trade studies.

The first element is the system requirements capture. A requirements traceability tool is essential to solidify the requirements, but the capture process is still something of an art that involves complex interactions among customers, system integrators, and specialty engineers. Asking the correct questions is an essential skill that system integrators must develop. In fact, interviewing techniques are critical to the success of new rule based systems.

An expert system might be able to mimic at least some of this interaction. The customer could speak into a machine that digitizes the voice input and responds back with spoken questions. The rules and knowledge based systems should mimic the interaction between a human expert and a system user. A typical interchange between humans might produce a session such as the following one.

Expert (E):	Do you have a computer systems problem?
Customer (C):	Yes, my embedded system costs too much to up-grade.
E:	Why do you say that?
C:	Well, we recently had to replace the data processor because it became obsolete. This caused a complete

	software rewrite because the code would not port. On top of that, the analysts predict that a similar upgrade is necessary in five years because of interface performance limitations.
E:	(Formulation of requirements such as use of widely available components, nonproprietary interfaces, software engineering environment, and similar features.)
E:	We can help you with that problem!

The electronic entries of this session are directly traceable to a particular set of customer concerns on a certain date. These concerns are likely to change over time into more emphasis on cost and schedule instead of engineering issues as the product release date nears. Once users receive the equipment, new engineering issues are likely to surface. The expert system can track these concerns over time and shift the emphasis or reformulate requirements as necessary. The requirements traceability tool can manage this interaction in much the same way that it does with human designers.

Available technology represents the second element of the process. This information comes from market surveys that engineering teams must conduct periodically. The current method involves periodic travel to research reviews, technical symposia, or conferences. The customers in this situation usually are engineers and product area managers from system integration teams, while the vendors are their counterparts in product specialty areas. A firm understanding of the capabilities and limitations of such products is crucial for a successful product design. However, much of the initial canvassing could be easily mechanized. An on-line system could perform at least the initial screening by receiving instructions from a senior systems designer to "find all the vendors of flat panel displays." Some refinement, such as the size, resolution, and operating environment could send an expert system off to do its work. The survey might involve accessing a vendor database via electronic modem. The expert system could identify promising candidates and pare this list to the top five vendors. The expert system could then use additional vendor data to produce a comparison list as a product for the senior systems designer. The designer might then instruct the expert system to request specific demonstrations, such as a particular display or graphics format. The product vendor could then supply a video tape of such a demonstration, followed by on site visits if the designer is sufficiently interested. This approach might dramatically reduce marketing and survey costs associated with both labor and travel expenses. The human interaction that is a key to a good working relationship need not be lost in this process. The expert system simply automates some of the more routine survey aspects.

The design engine consists of the strategies, rules, and algorithms presented earlier in this chapter, as well as other considerations. The earlier examples illustrate that it is not too difficult to automate the optimization of resource utilizations. However, these discussions did not relate how the utilizations of each function could be derived. A standard software interface

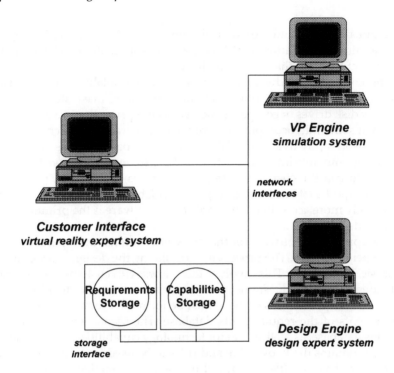

Figure 9-19   Example implementation of the design system.

to the requirements and capabilities databases is a key feature to tool integration. A standard interface, such as SQL, allows a design team to select among requirements and database tools from many vendors. In addition, this approach provides growth capability as new tools become available. Therefore, as in other cases discussed earlier, the individual tools exist, but successful integration of multiple tools into a design environment requires standard software interfaces.

The human interface is another consideration in the integration of a system. A state of the art sound or graphics system might require its own high performance host computer system. However, this capability is important to make the design system more accessible to design experts who might not be experienced in the use of databases or other specialized software.

A final consideration is the host system requirements in terms of storage and response time. A design implementation might partition these requirements onto multiple high performance work stations connected on a network, as shown in Figure 9-19. The first work station could contain the virtual reality software and hardware interfaces that allow the designer to interact with the databases. The interfaces might include the speech synthesis and voice recognition hardware and software, as well as the expert system software knowledge and rules base for requirements capture. Once the designer is satisfied with the requirements capture, simple instructions might allow this information to be entered into a product requirements database on a second work station. The second work station could also contain the capabilities database and the expert design system. The output files might be fed to a third work station, which serves as the simulation environment for the product baseline that serves as the VP.

The computer hardware for this system is available now. Hardware vendors provide work stations, network interfaces, mass storage devices (typically disk drives or compact disks) at modest prices that are lower in cost even than some microcomputers of previous generations. Much of the software is also readily available. This includes database systems with a standard, programmable interface such as SQL. Also, much of the virtual reality software to run the customer interface hardware is available. Finally, elements of expert systems have emerged from laboratories and are becoming increasingly more accessible. The application software is the primary area of software development.

The application software for the first work station is an expert system for requirements capture. The process of formulating the design rules is defined by classic AI methods. This involves interviews with customers and designers to identify typical questions or concerns, as well as rules to address such issues. The outputs of this system must be formatted to match the interface requirements of the requirements database. The second work station contains this database, as well as that for technology capabilities. The application software contains the knowledge and rules to be used in the design process. The product baseline files produced by the design engine must match the interface needs of the application code on the third work station. This consists of initialization files for configuring the simulations that comprise the VP. This third application is the simulation engine that offers a VP capability. This is the topic of the next chapter.

The power of virtual reality lies in its capability to establish a natural, intuitive, easy to learn input system. The AI technologies can automate many of the routine analytical tasks inherent in design. However, the design automation software inherently tends to become a point design for each type of product. An OOD approach to the design of these tools can maximize their generality and reuse.

References and notes

1. MIL-STD-499B "Systems Engineering Military Standard" (7 September 1993 draft), p. 6.
2. Som, Sukhamoy. "Node Assignment in Heterogeneous Computing," *NASA Contractor Report 4534*, August 1993.
3. Eager, Derek et al. Speedup Versus Efficiency in Parallel Systems, *IEEE Transactions on Computers*, Vol. 38, No. 3, March 1989, p. 408.

Projects

1. Consider the design rules of Section 9.2.
   (a) Extract some performance capabilities for at least two different types of single board computers. Vendors advertise systems of

this type in trade magazines, but you may need to contact the vendor for additional information. Include the memory size available to application software (not the total memory on the unit), interface types and performance, and processor type(s) and performance.

(b) Formulate a transfer matrix and resource utilizations for at least six software processes. Convenient numbers are acceptable as the next project revisits this task.

(c) Use the design rules of Section 9.2 to establish the number of processor modules needed assuming 50% reserves.

(d) Use the results of the last item to predict a system cost. Perform this estimate by adding the price of all of the processor cards to a mass storage unit (disk or compact disk drive). Include also the cost of a power supply and enclosure. (Hint: If you pick a processor module with a standard size and interfaces, you will have many choices. Experiment with this!)

(e) How much does your design weigh?

2. Functional requirements can be quite difficult to characterize accurately. This is often especially true of performance requirements. Consider how one might derive such estimates.

Modern software is written in a modular fashion to both simplify initial testing and control the complexity of the integration effort. Suppose that the software manager insists that a unit may not contain more than 500 source lines of code.

(a) Pick some software application, such as an embedded control system. Decompose this into top level functions. How many top level functions are there? Decompose at least one of the functions into units. A simple example, such as an interface is sufficient. The "brute force" approach to this problem is to develop prototype code. How many units are there? How many source lines of code do you estimate?

(b) Assume that the compilation system generates, on average, four assembly language instructions for each source line of code. How many instructions are involved in your application? How many MIPS can your single board computer provide? How much time is required to execute your application? How fast must this function provide an input to your application? What is the processor utilization of this function?

(c) Assume that each instruction requires 4 bytes of storage. How much storage does your application require for code? Assume that the data requires an additional 50% storage. How much total storage does your application require? What is the memory utilization of this function for the single board computer discussed in the first project?

(d) Ordinarily, analysts do not generate requirements ab initio as this project example suggests, but also use performance benchmarks and historical data from other software applications to supplement these estimates. Discuss how such data might be modeled. (Hint: Reconsider Equation 9.3; is it possible to build a more general model?)

3. For this project, you may wish to reconsider the results from the Chapter 6 projects.

(a) Identify and price a high performance work station.

(b) Identify and price the following software:
   - VHDL simulator and compilation system,
   - requirements management tools (or database system),
   - database system capable of storing the technology capabilities. Assume at least 15 fields per record and at least 500 records.

(c) Suppose that the two applications described in Section 9.4 each require four persons 6 months to develop. What is the cost of this application software to develop? Assume that software maintenance requires two full-time engineers. What is the total amount of investment required after 5 years?

(d) Suppose that the VP system increases design speed ten times. If a typical design costs $750,000, how long before this system reaches the "break even" point?

4. Reconsider the design of project 1. Suppose that the utilization requirements increase 10% per year, for each of five years.

(a) Discuss how your design can grow to meet these requirements.

(b) Compare the cost of the initial system to that of the final system (assume constant dollar value per year).

(c) Compare the reliability of the initial system to that of the final system.

(d) Assume that the following trends in new single board computer designs is constant:
   - Memory capacity doubles every 2 years
   - Processor throughput doubles every 5 years
   - Interface speeds double every 5 years.

   Using the cost estimate of 4(b), at what point is it less expensive to buy new computers instead of increasing the performance of the existing design? Assume that the software can be ported without change.

5. Consider the interface to the requirements engine.

(a) Discuss the general nature of a natural, intuitive input system. Translate this discussion into interface requirements.

(b) Which virtual reality technologies might be applicable for this interface? Be specific by considering both the technology and the interface needs.

(c) Consider some components of the expert system, especially the application. What elements might apply to all three engines (design, prototyping, manufacturing)? How might OOD be applicable to these considerations?

chapter ten

Prototype evaluation

A frustrating, but common, dilemma for a system designer is the impossibility of attracting financing, beta test sites, or software developers without some type of working prototype. Construction of a fully functional prototype can be daunting because of the expense. However, automated tools might offer another way to demonstrate features of a design without actually building hardware and integrating software.

Figure 10-1 demonstrates a possible automated design process. Each arrow represents some type of tool that accepts the results from one type of model and generates new software for the next stage. As mentioned previously in Section 5.2 (see Figure 5-7), the heart of this approach is the VHDL behavioral simulation capability. Automated tools can generate such models from graphical inputs, such as STDs, class and object diagrams, and timing diagrams. Additional tools can synthesize physical descriptions and Netlists from component behavioral models. These are also the basis of the graphical description for computer modules and data buses. It is also possible to produce system simulations from track files built from execution of the module descriptions. A comprehensive system level test is possible with the final product, which can be a real-time simulation.

The VP method can allow an earlier look at areas such as:

- System requirements validation and reuse
- Problem solving methods for simulation software integration that can serve as models for actual product integration
- Adaptability of the design for changes in requirements that might be related to upgrades.

Requirements validation is always an important but complex task. Often this requires construction of a working prototype to gain full confidence that the contract documentation contains technically correct requirements. The VP can be an important milestone in this process, which, in some cases, can obviate the need to build brassboards. Furthermore, the VP process relates these requirements to abstract behavioral models that can simplify reuse of them in other system designs. This may eliminate duplication of costly trade studies.

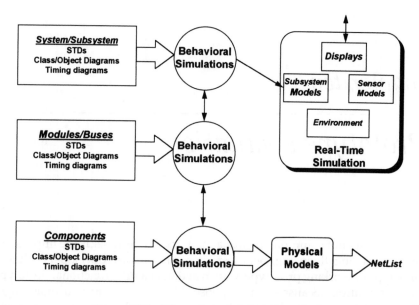

Figure 10-1   The automated design method.

Modern design methods can inherently simplify testing of software because the final code will feature a high degree of modularity and the proper visibility of data structures. OOD has the additional benefit of providing a model sufficiently abstract as to maximize requirements reuse. The automated VP design process also helps in this regard because system level test vectors and outputs can be synthesized from module and component test code in the same manner that architectures can be synthesized from behavioral simulations.

The cost effectiveness of the automated generation of a VP varies with the design application. For example, an environment to test avionics might cost $1.5 million. This can be traded off against flight test hours ($10,000 or more per hour) and the cost of fabricating a brassboard unit. The cost of the latter is often very high, especially if little applicable off the shelf equipment is available. Therefore, the break-even point for this facility might be as low as two to three subsystem prototypes. On the other hand, consider a desktop computer design. One might argue that the market price is around $2,000, so extensive simulation may not be cost effective. However, a hardware prototype would cost many times this price. A more relevant question is the type of testing VP offers for a desktop computer. The system level functional and performance requirements are (apologies to the designers) relatively uncomplicated. Unit cost is the real driver for such systems.

Another value added by the automated VP process is insight into system integration. The VP provides a check of the design assumptions that the component and module designers build into their baselines. The VP also offers the first look at system performance, interfaces, and functionality for software designers. This might be studied as part of an expert system, as Section 5.1.3 describes. Experience teaches that the software engineers must gain this insight early to impact the software requirements formulation. A hardware prototype may not be available until first generation software is on line, which often forces managers to make expensive decisions about changing hardware or software requirements as problems emerge (and problems always emerge during integration.) In fact, this is one aspect of the 1980s software crisis for complex systems. The hardware design was often largely complete before software requirements could be validated. This sometimes resulted in embarrassing processor performance, I/O capacity, and memory size deficiencies that slowed or killed projects. This suggests that the automated generation of a VP complies with the spirit of the following system design rule:

> *A validated product baseline must be based on equally*
> *mature requirements for hardware and software.*

That is, the software requirements must be in the critical path of the system design. For those who doubt that major programs ever violate this rule, simply study the schedules for the various hardware and software major reviews. The time difference should be alarming to the alert reader, often a difference of more than a year!

Finally, reconsider the iterative process when a designer encounters problems with a prototype. A defective hardware prototype might require very expensive and time consuming modifications such as new component logic, changes to the placement of components or traces on the module, changes to power on start-up sequencing, changes to error recovery methods. The integrated circuit automated design process suggests that the VP approach is the most cost effective method of dealing with such problems.

The examples in this chapter illustrate the VP approach to system design. Two very different computer networks are the subjects for these examples. However, the discussion begins with an overview of the simulation tools and process elements essential to successful virtual prototyping.

10.1 The VP simulation engine

The design engine of Chapter 8 can formulate baseline designs and ranks of candidates. However, the basis of this ranking is a comparison of technology capabilities against component or unit requirements. At some point, the analysts must integrate the building blocks into a prototype that proves the system meets the top level requirements. The behavior and performance of the units may seriously degrade in a system environment (see the discussion in Section 9.1.2.5), so the capability to explore these effects early in the process represents an important risk reduction.

Rapid prototyping is a widely used term, but the concept is key to risk management. An early look at the behavior of a prototype can help to validate requirements at all levels. These prototypes can also serve to quantify manufacturing costs and time, acceptance test procedures, and maintenance processes. The initial phase of the prototyping serves as a prelude to system integration in much the same manner that successful completion of software unit testing is the basis for system test procedures.

Many rapid prototyping processes focus on manufacturing and assembly teams that are highly trained and adaptable in quickly fabricating a working model of a design. However, the increasing performance and decreasing cost of simulation systems offers another intriguing possibility. The same bus, processor, and interface simulations that are so valuable as evaluation tools in trade studies can provide many of the same capabilities as a hardware model. This discussion explores some of the uses of the hardware models and the capability of simulations to rapidly evaluate design baselines for these purposes.

As suggested earlier, hardware prototypes serve a variety of purposes. These include validation of performance requirements as well as proof of the design concept to meet such needs. In addition, virtual reality systems are useful in exploring manufacturing and maintenance concepts.

System performance is the focus of this topic, but consider for the moment the issues of manufacturing and maintenance. Finite element simulations can capture considerable detail about the physical nature of the system and its environment. A graphics work station can reproduce automatic assembly procedures, such as component placement on circuit boards, wiring harness assembly, or even system qualification testing. The system can be modeled as a three dimensional grid whose node points contain values that represent some physical characteristic, such as color, temperature, or the forces applied at this point. The analyst can then allow the model to interact with its environment to determine its limits. This might be the operating temperatures, mechanical vibration, mechanical clearances available during the assembly cycle as parts are added, or a variety of other factors. Although many tools of this type exist, a more integrated approach to the software tools could increase reuse and decrease cost.

A comparable approach to performance analysis is not widely practiced and offers some interesting challenges. However, a clearer conceptualization of the system and its testing are requisite to an exploration of these problems. A computer system includes both hardware and software, so a correct behavioral simulation must account for both aspects. A very general processing model that practitioners of SA use is control/input/process/output (CIPO).[1] While the distinction between control and processing may not always be cleanly drawn for a real-time system, this model is a good starting point. The implementation of this model requires that the analysts identify I/O from large sections of software, the function that this software provides, and the control and timing of operations that trigger this function. The utilization tables and node assignment rules of Chapter 8 partially answer these general questions, but the behavior is more complex in a system environment in contrast to single board computers operating in isolation.

A model of communication within the system is an important feature of a VP. However, the performance characteristics are much more difficult to capture for the full system than the I/O utilizations of the last chapter suggest. The concerns of an end user most likely revolve around *end to end latency*, which is the amount of time between the transmission of a message and its receipt by the destination function. The calculation of this time

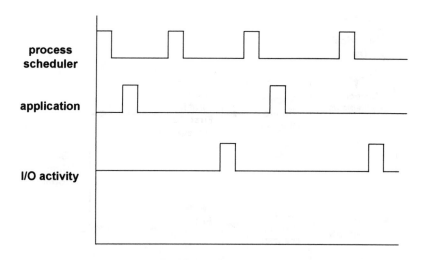

Figure 10-2   Sample software timing diagram.

involves a detailed tracing of the message through the various processor, memory, and interface systems. However, even this calculation is incomplete since software functions must compete for access to these resources. This competition is an essential characteristic of the system model.

Software execution time is another characteristic of system performance that a VP must include. As with end to end latency vs. I/O utilization, calculation of this parameter involves much more than a time estimate for a single application, as in Chapter 9. Instead, the software on a node must compete for resources such as memory or processor(s). Scheduling algorithms determine the winner of this competition, but luckily there are only a few commonly used algorithms.[2] The two examples that follow are based on an *on demand* approach and *message passing communication*. The distinction lies in statistical vs. scheduled requests, but the operation of the software produces a time line, such as the one shown in Figure 10-2. The process scheduler constitutes the control operation of the CIPO model, a separate I/O time line covers both the input and output of the model, while the application time line represents processing. A time line of this type must be constructed for each node that includes all such operations. Simulators are already in wide use for this type of analysis, although few standards prevail.

The simulation engine of the VP must contain a number of behavioral simulations for implementations and types of both serial and parallel communication protocols, as well as processor and memory combinations. The VP environment must also provide a method to extract the key features of the top level software and system functions from the requirements inputs. The outputs of the design stage and historical data must form the basis of performance and interface definitions for these functions. From these inputs, the timing diagrams for all of the processes assigned to a node can be formulated as shown in Figure 10-3. This portion of the model represents the processor and memory interaction with software, or *host side* behavior of activity on the

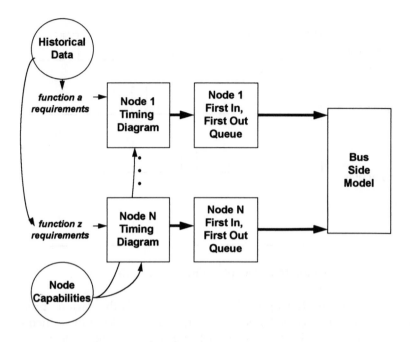

Figure 10-3   The VP elements.

node. The *bus side* model consists of the timing and protocol features of the bus type. In actual implementations, an I/O controller and dedicated memory buffer the host side from the bus side. This is easy to model as a first in, first out queue between the models for the host and bus sides as Section 9.1.2.6 suggests. The final component of the simulation is the executive.

The simulation executive provides the resources to schedule simulation events. This might be approached as a nonreal-time control loop that updates at a fixed time increment. At every tick of this clock, the execution chains for each node and the bus system increment. The timing patterns repeat after some amount of time, which may be different for each node, after which the sequence begins again. If the software is an on demand system instead of scheduled processes, a statistical model might generate the software events. Regardless, as the I/O queues fill, the nodes begin to compete for access to the communication channel. The simulation should contain the equivalent of a debugger, which allows the analyst to trace a message through the system to establish end to end latencies. For scheduled processes, the worst case, best case, and average case timings are often the results of interest. For an on demand system, the end to end latency vs. the request rate is usually the behavior of interest.

Once again, much, if not all, of this technology exists now. As discussed previously, one example of this design process is a virtual silicon foundry whose function is to design and test microcircuits using simulations. This same type of technology can offer powerful insight into system behavior. The subsequent examples provide more details of this type of system.

10.2 Scheduled processes example

Section 9.1.2.6 presents an overview of the VP approach to performance modeling. This discussion introduces the idea of software scheduling. This important aspect of performance modeling is the topic for this exposition.

In this example, a predetermined schedule for several software processes establishes the control flow. Figures 10-9 and 10-10 (case 2) provide the basis of a good example. This involves 16 processes assigned to 7 nodes, as shown in Table 10-1 (this is the same example that Section 9.2 uses). For simplicity, assume that the base tick rate for each node, or *major frame*, is 0.05 seconds (20 Hz), but these are not synchronized due to differences in start up times and clock drifts. The time increment for the simulation executive must be the smallest of the times involved, but to update faster is to simply waste processing time.

The times of execution are derived from the requirements inputs to the design engine which are the origins of the utilizations. The numbers of this example are artificially formulated in this example. The time per process is simply the utilization percentage of the major frame. Table 10-1 shows the results of this estimate. The order in which the processes attempt to execute is also important. Ordinarily, this is determined by a priority scheme, so for this example, simply assume that a lower process identification number represents higher priority. This forces the processes to execute in numerical order. Two other factors are important. The first is the control overhead of the scheduling executive. A proper utilization estimate should already contain this overhead, but the analyzer might also base it on a fixed percentage times the number of processes for each node. The second factor is I/O impact.

The I/O impact can be somewhat more difficult to establish. The access competition of the processes for the I/O queue should be measured as part of the simulation. However, there is still some overhead associated with this competition and the configuration of the I/O hardware controller and dedicated memory. A proper utilization estimate should also include this overhead, but for a new design the analyst may not have values for this transaction. In such cases, analysts often model the overhead as a fixed percentage of the total execution time for each process.

Table 10-1 Processes and Nodes for the Example

Node	Processes	% Node utilization	Time (ms) per process
1	1,2	25	8.25
2	3,4	45	11.25
3	5,12	45	11.25
4	6,13	45	11.25
5	7,8,9,10	50	6.25
6	11,14	35	8.75
7	15,16	55	13.75

Now the I/O utilizations must be turned into transmission times. To do so requires the understanding that utilization is often quoted as the ratio:

%util. = 100% * (number of bytes to transfer) * (update rate in Hz)/bandwidth.

(10.1)

In this expression, the number of bytes includes protocol overhead, such as header or ending delimiters, as well as data words. The update rate is the scheduled frequency of transmission, with 20 Hz being a common choice. Bandwidth represents some measured or calculated burst rate in MB/s.

The transfer time depends on the bus implementation. As an example, consider a synchronous, parallel bus that transfers four bytes per clock cycle. The update time for a message is:

$$\Delta t = \text{(number of bytes to transfer)/(signal rate in B/Hz).} \quad (10.2)$$

As a result, the time relates to the utilization by:

$$\Delta t = [\text{bandwidth} * \%\text{util.}/100]/[(\text{update rate in Hz}) *(\text{signal rate in B/Hz})].$$

(10.3)

For this example, suppose that the bandwidth is 4 MB/s, the update rate 20 Hz, and the signal frequency 20 MHz, then

$$\Delta t = [\%\text{util.}/100]*(0.010 \text{ s}). \quad (10.4)$$

This represents the actual time on the bus for the message, not the end to end latency since it does not include host side transfers. It could include competition (*bus vies*) if those control words are included in the message total byte count. However, this allows one to neatly break the transfers into 50 ms major frame blocks using the I/O utilization matrix. If one begins with the first row and sums utilizations across to break them into blocks of 100%, then 5 such blocks represent 50 ms. The sum of I/O utilizations in the transfer matrix is 2175 (refer to Section 9.2), which represents 0.2175 seconds to transfer the entire message list. One must subtract from this the contributions of processes assigned to the same node, whose I/O is therefore host side only. This sum is 185, making the total transfer time 0.190 seconds. If transmitted in priority order as suggested earlier, this is exactly how long it takes to complete the list and circle back to the first message on the list for the bus operations.

The bus list transmits as soon as the controllers can fetch the data. In this case, the system is I/O bound because the software processes can complete more than once per transfer cycle. The artificial data is the source of this problem as it is unlikely that processes would have such high I/O utilizations. However, this is a clear signal that the bus selection is inappropriate since scheduled events dictate that all activity complete in one major frame. The I/O activity requires four major frames to complete in this case.

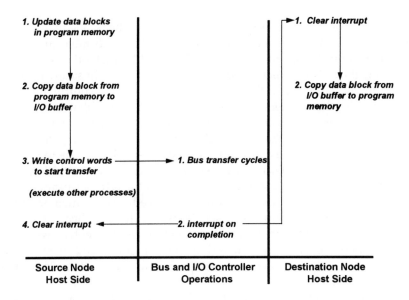

Figure 10-4   Message transfer cycles.

One of the host side delays involves scheduling the software on both the source and destination that configure and execute the transfers. One highly efficient method of host side control is *memory mapped I/O,* in which the registers of the I/O controller are directly attached to memory locations accessible to the data processor. This permits the software developer to simply perform memory writes to initiate a transfer. On the source node, the I/O controller sets a hardware interrupt that signals the software that the bus transfer is completed. On the destination node, another interrupt signals reception of the message to the application software. The I/O process on the destination node must then move the data out of I/O memory into program memory. Figure 10-4 shows this process and the associated timelines.

The cycles of Figure 10-4 show the entire sequence involved in a calculation of end to end latency. This consists of hardware and software configuration on the source node of the message, bus operations, completing with reception and clean up on the destination node. The interrupt sequence in this figure is greatly simplified for clarity.

Although the I/O bound case is somewhat pathological, this figure suggests quite generally that software overhead can swamp the bus transfer times. Two general contributions are the software process *scheduling delays* and the RTS performance.

If the major frames are not aligned, the scheduling delay can be as large as one major frame. In some systems, though, the alignment of major frames is enforceable using a low latency *global time message* that a controller node broadcasts. Special hardware is required for this timer. However, unless the system designer tinkers with schedules, it is still quite likely that the destination processor could copy the data block in the next minor frame.

These factors are quite important in control or other embedded processing applications because message times must be predictable in order to assure that control loops function properly. An older method of solving this problem is the *critical timing loop* in which the designer physically measures the end to end latency and then hard codes this into the control algorithm. Unfortunately, this renders the software a point design that must be discarded if an upgrade introduces faster hardware into the system. A *global time distribution* approach involves establishing synchronization points among the nodes using some very low latency mechanism. This tends to be much more portable because it schedules events instead of directly timing them.

The second issue in host delay calculation is the RTS performance. This is simply the software overhead involved in procedure calls, memory writes, and controller configuration for each transfer. This overhead makes the theoretical transfer rate, including bus competition and other hardware overhead, unattainable. A common derating factor in embedded systems using tasking is to assume an achievable capacity no higher than 25% of the theoretical estimate.[3]

This example illustrates some host side and bus side delay calculations for transfers involving scheduled processes. Different elements of the behavioral simulations can be the basis for estimates. These include models for interaction of processors and memory, software schedules, and bus hardware operation. However, these individual elements are insufficient to estimate the end to end latency that is the system level requirement. Instead, the behavioral simulations must execute in a system environment since as a design incorporates more nodes in its network, both host side and bus side delays increase. However, it is quite reasonable to build up the system model from elemental simulations that describe the behavior of a single aspect. This approach is equivalent to software integration, which begins with testing of low level units and culminates in the stimulation of all units interacting. This also serves as a metaphor for the process of integration of the actual hardware and software. Consequently, the models that comprise the VP can serve as requirements validation and proof of concept for the design baseline selection.

10.3 *Statistical processes*

The previous example considered scheduled processes whose predictability of timing is important. In many applications, though, service requests are quite random. For example, a multi-user work station might have service peaks, but a user could log on and start a process at any time. For such applications, the response time of such processes vs. the number of users is a crucial design consideration. This problem is the basis of the examples for this chapter. Basic examples illustrate key concepts and are gradually generalized to those of statistically based simulations.

The scheduling algorithms for systems of this type are based on the concept of *on demand scheduling*. This means that service requests for various system resources arrive randomly so that the system must have sufficient

Table 10-2 Process Scheduling
Algorithm

Process burst rate	Tick = 1	Tick=3
1	1 X X X X X	1 X X
2	X 1 1 X X X	X 1 X
3	X X X 1 1 1	X X 1

Note: 1 = executing, arbitrary time units; X = not executing, arbitrary time units.

performance to schedule events asynchronously. The worst case scenario in this type of architecture occurs when a user process must wait for the transmission to complete its execution. That is, the transmission is a *blocking send*. The general behavior of such a system is that as more users log on and start processes, the slower the system responds. At some point, the response times are slow enough to not just be distracting, but to decrease productivity.

In practical terms, an operating system for this type of computer system schedules from priority queues. User processes, though, are the focus of this example and generally run at the same priority. Within a priority group, subpriorities might be based on a number of algorithms:

- First come, first served
- Non-preemptive, shortest job first
- Preemptive, shortest job first.[4]

The first come, first served scheduler simply selects the oldest process in the queue to execute. On the other hand, shortest job first allows the process whose burst rate is smallest to run first. Burst rate is the estimate of execution time when the process runs in isolation.

The problem with the first two algorithms can be *fairness*. This means that the average amount of execution time per process is constant for the system. Preemption is a mechanism that allows the selected process to execute for one *operating system time tick*. The operating system monitors a real time clock and loads a new process at periodic rates.

Consider now a simple example. Suppose that three processes have burst rates of one, two, and three time units and execute a required single execution to complete. Three cases of time ticks of one, two, and three time units in the preemptive, shortest job first algorithm are shown in Table 10-2. The finer resolution (tick = 1) is more efficient because it allows for less overall amount of time (six time units for case 1, nine time units for case 2).

A more complicated example is not too difficult to imagine. Suppose that a number of users on a system are using a common software resource, such as a text editor. In this case, users may edit for many minutes or hours, so the previous example must be generalized to processes that do not terminate. Table 10-3 illustrates this case. A process that has been executed is placed at the end of the ready queue. The queue therefore completely cycles after a

Table 10-3 Continuous Process Scheduling
Algorithm

Process burst rate	Tick = 1	Tick = 3
1	1 X X 1 X X 1 X X	1 X X
2	X 1 X X 1 X X 1 X	X 1 X
3	X X 1 X X 1 X X 1	X X 1

Note: 1 = executing, arbitrary time units; X = not executing, arbitrary time units.

number of time ticks equal to the total number of processes. The *refresh time* is the number of time ticks needed between executions of a process. In both cases, this is two time ticks. However, as the refresh time increases with the number of processes in the ready queue, each process cycles to the end of the queue after it completes. This is called a *circular queue*.

Next suppose that each process needs multiple resources to execute. This might involve disk drives, user terminals, tape drives, or other devices. In this case, each device can have a process queue. This is an example of *multiple resource queues*. Processes from these queues can be scheduled preemptively by the controller for each device using an algorithm such as shortest job first. Since the access speed of the devices is so different, the queues are likely to have different tick values, queue depths, and other key features. In addition, successful completion of the process probably requires that the resource usage be in some sequence, not random. Therefore, the process at the head of a device queue might be *blocked* from execution because it must wait for a process in another queue to execute. There are two methods to handle this, *interlocked* or *non-interlocked*. An interlocked queue systems allows blocked processes that reach the top of the queue to wait while a non-interlocked queue cycles blocked processes. Table 10-4 shows simple examples of both types for two queues and three processes. The first processes that must execute are processes 1 and 3 for queue 1 and process 2 for queue 2. The queue entry of the process is equal to its burst rate in time ticks. Also, since processes do not migrate, subscripts that indicate the queue to which the process resource element is assigned are omitted.

In the table, notice in the non-interlocked case that process 2 blocks and cycles so that the execution order is 132. Queue 1 returns to its original state in four ticks in the non-interlocked situation. In the interlocked case, queue 1 must wait on resource element 2 in queue 2. In this circumstance, queue 1 returns to its original state after eight ticks. Queue 2 is the same in each case and cycles in six ticks.

This simple example illustrates the power of predictive scheduling. The trends are not general because non-interlocked queues do not always cycle faster than interlocked queues. However, a thorough understanding of the timings and resource needs of the processes allow the designer to select the optimal approach to reduce cycle time. In the example, user response time is directly related to the cycle time of a process through both queues. This is

Table 10-4 Multiple Queue Example

	Queue 1 with interlocks	Queue 1 no interlocks	Queue 2
t = 0	1	1	1
	2	2	2
	3	3	3
t = 1	2 (blocked)	3	2
	3	1	3
	1	2	1
t = 2	2 (blocked)	3	2
	3	1	3
	1	2	1
t = 3	2	3	3
	3	1	1
	1	2	2
t = 4	2	1	3
	3	2	1
	1	3	2
t = 5	3	2	3
	1	3	1
	2	1	2
t = 6	3	2	1
	1	3	2
	2	1	3
t = 7	3	3	2
	1	1	3
	2	2	1
t = 8	1	3	2
	2	1	3
	3	2	1

Note: Process 2 in the first column must wait on process 2 to complete in the third column.

similar in nature to the instruction pipelines of a data processor. Typical variables in the simulation of such systems include queue depth (number of entries permitted), priorities, and interlocks. The priorities relate to the urgency of execution and can resolve blockages. For instruction pipelines, for example, instruction might be classed as privileged or not privileged, with a privileged instruction receiving priority in the case of a blockage.

One final issue is *deadlocks,* which occur when a high priority event must wait on a low priority event to run, but the low priority event is blocked. The interlocked case shows an example if queue is the higher priority queue. At t = 1, process 2 is at the head of queue 1 and is blocked while process 2 is also at the head of queue 2. The blockage is unresolvable because queue 1 has the priority, but the process at the head of the queue is blocked. There is no choice in this case except to break the deadlock by *flushing the queue.* During design, a resource allocation graph helps to identify deadlocks.[5]

The next step toward a more general model is to postulate a statistical arrival of software processes at the queues. In addition, the burst rates might be randomly distributed as well. Queuing theory is the branch of statistics that addresses this problem domain. A typical arrival process of such models is based on the Poisson probability density function. However, such problems quickly become intractable and computer generated, numerical solutions are necessary.

Many of the statistical simulation tools rely on a graphical method called stochastic Petri nets. In very rough terms, this approach is similar to other graphical methods described earlier. Imagine the STD for a system. Assign a transit time and probability to each transition event of the STD. If the transit times are also randomly distributed, this diagram captures the essence of a stochastic Petri net. Simulations of this type can be very complex, but also quite successful in describing the operation of a network.

To describe such simulations in any more detail would be too distracting. However, since these simulations can accurately describe the operation of computer networks, the reader should ponder their place in the realm of automated design tools. STDs continue to be the entry point for many of the automated tools and the same is true for statistical simulations. A simulation executive might use an STD to configure and run the simulation. Beyond this, the STD might also be the basis of VHDL behavioral simulation. These two capabilities are synergistic because the VHDL simulation might generate the timing parameters that are the inputs to the statistical simulator. The resulting environment is a powerful design validation system with an intuitive, graphical interface.

The simulation tools based on statistical methods cover an important segment of the design problem domain. A fully integrated environment can be an element associated with the generation and testing of a VP. This capability allows rapid evaluation of designs, requirements validation, and documentation. Such tools are now available for popular work station hosts. The current popularity of such tools and projections of their capability are likely to force even the most conventional design engineers to learn such methods.

10.4 The VP engine

The previous discussions of this chapter point out some problem areas and simulation requirements for a VP engine. In addition, Section 9.1.2.6 describes how the VP serves as a bridging activity between design and manufacturing. Nonetheless, many of the issues dealing with computer environments are common to all three phases. This discussion provides a more complete description of the VP engine, but much of the discussion is general.

The overall goal of the *virtual enterprise* is to produce a highly integrated system so that outputs from one stage of the process merge seamlessly into the next stage. The phases are:

- Design (development of a product baseline)
- Prototyping (proof of design, proof of manufacturability)
- Manufacturing (production).

To build such a system for all possible product areas, though, is impractical. The designer must begin the VP engine requirements formulation by carefully defining the types of problems that the system must solve, also known as the *target problem domain*. This allows the VP integrator to identify design rules, algorithms, and interfaces. Within such a domain, one must break the analysis into simple stages to form an analysis pipeline. This capability to merge results in this way depends on the automated tools used for each phase, such as requirements formulation and tracking, STD entry and simulation, and VHDL considerations. The final issue is validation of the results.

Previous discussions point to several problem domains. Although this book focuses primarily on computer system design, the methods are applicable to any complex design including highway traffic flow in civil engineering, pedestrian traffic flow simulation in architecture, ground based telecommunications, and a panoply of others. The methods are general but usually require tailoring or shifts in emphasis for a particular application. Within computer system design there are also classes of problems, such as system network bridging and information flow, core computing needs, and the various needs of specialized, embedded computing. Once again, the VP engine designer must refine the definition of the problem domain so as to avoid excessive performance needs and concomitant high costs.

Another factor is merging results. This may seem obvious, but the VP engine integrator should avoid special purpose, custom tools because of potentially high development and maintenance costs. Therefore, one aspect of tool selection is the ability to import and export files in some standard format.

Validation of the results is also a concern. The key issue is how one might test the VP software to verify that it is operating correctly and represents a high fidelity model of the problem class. This can be an especially annoying problem with expert systems for which it may be difficult to reach all branches of the logic flow with test inputs. One important but little used method is comparison to hardware results. During initial design, design configurations can be run through the VP engine for which hardware is available. The predictions of the VP engine can then be tested on the actual hardware. This also allows the VP designer to know where the limits of experience of the VP engine are located. This is equivalent to a manager understanding the experience of a design engineer on a development team to assure that the engineer is not assigned to unsuitable tasks.

A final issue on validation is the certification process. This will vary depending on the customer, but some type of formal reviews and final authority to proceed are very common approaches in managing the contractual relationship. Since this often means that the designer is paid incrementally after completing each major review, the VP engine must be capable of producing results to support such reviews, as well as to provide demonstrations for participants at reviews. This can allow the design team to quickly resolve action items stemming from such reviews.

Figure 10-5 revisits elements of the design and VP processes. The system requirements represent the core of the enterprise. Automated tools to capture

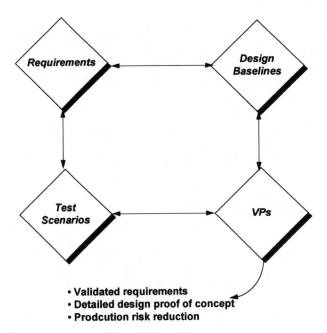

Figure 10-5   The VP cycle.

and maintain requirements, as discussed in Chapter 8, can feed them in some
standard electronic format to other analytical engines. The design engine that
produces one or more baselines is the topic of Chapter 9. Test scenarios are
essentially simulated functional tests that might otherwise be performed on
the hardware prototype. The VP engine consists of a library of simulation
elements that the design engine can configure. For computer design, this may
be various bus and processor simulations. The VP engine must configure
these elements to extract the results needed to answer the test scenarios.

The previous chapter discussed a rule based design method that can
simulate how an expert in computer systems design might formulate a
baseline. The rules for a VP engine deserve equal attention. The VP engine
serves two main purposes. The first is virtual trade studies that prove the
validity of the design. The second purpose is synthesis that transforms a
behavioral description of the product into a physical description. The two
key elements that feed the VP engine for trade studies are the design baseline
and the test scenarios. The synthesis component draws upon the product
baseline and hardware and software technology descriptions.

10.4.1 Virtual trade studies

The dividing line between the design and VP engines is domain specific. For
some problem types, it may be relatively easy to add tools that allow an
engineering team to perform trade studies directly with the design engine.
Other problems may be too complex, or require too much computing power,
to adopt this approach. For this reason, the following discussion retains the

distinction between these two elements although they might not be distinguishable for a particular implementation.

In order to configure a design within the VP, the design engine must establish which VP library elements are needed and their relationships. Chapter 9 discusses how electronic files that capture the design can be the basis of configuring the simulation elements for the VP. The examples of this discussion revisit configuring the simulation environment. In addition, the significance and use of the test scenario is a crucial element of virtual trade studies.

Manual entry is simpler with a point and drag graphics interface, but an electronic file output might be more efficient. For the latter approach, a standard interface language is likely to be effective. This language might resemble that of Figure 10-6 for a dual redundant channel that connects three modules. In this example, BUS_A and BUS_B are the names of the channels in the simulation and the type 32_bit_parallel_bus is a simulation library element. The remainder consists of the module names and references to library elements for modules of the indicated kinds.

This definition is derived from the output of the design engine of Chapter 8 that assigns software to nodes and determines module counts and types. These outputs do not reflect a particular interface or module, but simply characterize the requirements. The VP engine executive assembles behavioral simulations of actual products into a system simulation. This process involves matching library elements against the stated performance and functional indices that the design engine provides.

The simulation scenarios accept the I/O load matrix and other performance information from the design engine and requirements phase. The output of this stage is a *I/O message mix* and *process schedule tables*.

The I/O mix is a byte count and update rate between each pair of software process pairs. This is quite general as the control words surface within the simulation. However, as the earlier discussion describes, it may be necessary to return to first principles instead of directly using I/O utilizations from the design engine. This happens often with manual design entry as well.

The process schedule table is the order and rate of execution of software on each node. This may be based on priority preemptive scheduling with fixed, hard deadlines. Ordinarily, such analysis assumes that each process runs to completion during each operating system tick. Task suspensions, rendezvous, deadlocks, and other phenomena might be modeled in parallel by those performing the software design.

To cement these ideas, consider a new computer system design with two candidate parallel buses and two candidate data processors. The goal is to find the best combination of processor and buses to meet the requirements. This involves four combinations of elements.

The testing is where candidate rankings emerge. The virtual trade study test scenarios should emulate the acceptance test procedures that the customer uses to buy off a new product. Each requirement in the product system specification must have a corresponding test element in order to be enforce-

```
network

        BUS_A : 32_bit_parallel_bus

    connects

                MODULE_1 : 32_bit_data_processor ;

        to

                MODULE_2 : 32_bit_data_processor ;

        to

                MODULE_3 : network_serial_channel ;

        end;
end network ;

network

        BUS_B : 32_bit_parallel_bus

    connects

                MODULE_1 : 32_bit_data_processor ;

        to

                MODULE_2 : 32_bit_data_processor ;

        to

                MODULE_3 : network_serial_channel ;

        end;
end network ;
```

Figure 10-6   VP interface code example.

able by the customer. The formulation of such tests involves application of some input to the system in order to measure its behavior and resultant outputs. The input might be digital data words at a computer interface or could be an environmental factor such as temperature. The test plan must also avoid unrealistically artificial scenarios since many of the factors are coupled in complex ways.

The test plan is an output from the design process. Some factors, such as the environmental requirements, are likely to require a physical model of the system. However, a behavioral model may be sufficient for performance and interface testing.

Consider an example of such elements. The detailed requirements may seem rather complex at first, such as: "the system shall, upon failure of the primary bus, halt operation on the primary interface and initiate operation on the secondary bus within 10 ms."

The test scenario code must further decompose this into low level performance requirements such as:

- The worst case time needed for a module to detect that the primary bus has failed
- The worst case time needed to signal within the network that a switch is needed
- The worst case time needed for a module to switch to bus B
- The worst case time for the previous master module to become master of bus B and begin transmission.

It may be too inefficient to derive the test plan and associated test vectors using simulation methods entirely. However, reuse of previously validated test scenarios can also represent a cost savings. Instead of deriving these detailed requirements, the test scenario generator may rely heavily on *document templates*. The reader might compare this method to the object and class definitions from OOD.

In tools that provide automatic document generators, a template provides the structure of the document, but not the contents. Instead, the contents flow from some other aspect of the tool, such as an STD or the automated system that captures requirements. The test scenario might function in the same manner for production of detailed requirements using templates for the system, module, and bus specifications.

Therefore, in the example with two processors and two bus candidates, the combinations of processor types and buses is of the form {AA, AB, BA, BB}. Usually, a processor will couple more naturally to one type of bus system to another. This can be shown, not simply in performance, but in the amount of support logic needed to connect the processor and its memory system to the bus control interface. These are design issues and may be inputs from the design engine, but the VP engine should be able to characterize cost and schedule issues of such designs, not just performance. In addition, the VP engine must discard designs that cannot meet the requirements and rank the remaining designs in some preferential order to establish a recommendation.

Figure 10-7 shows some simple bus and processor types that might be the basis of an evaluation. Once again, the module design feeds into the VP engine, but there are still performance, cost, and functionality considerations. The difference between the two processor types lies in the memory interfaces. Processor A provides separate memory channels for program execution and I/O. Processor B uses a memory hierarchy involving primary (internal) and secondary (external) caches and a local memory. This type of design requires extensive logic to provide cache memory data consistency, but can also offer low cost and high performance with a suitable design. The distinction between the buses is the memory and control interface. Controller A is formulated for memory mapped I/O, which means that the application software can modify control registers and transfer data blocks by writes to dedicated memory locations. Controller B functions as a coprocessor to the

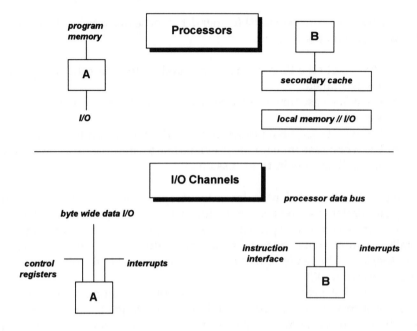

Figure 10-7   Bus and processor types for the example.

data processor. This controller has both an instruction stream and a data interface.

The module simulations must couple the bus controllers, processors, and memory systems. Once these models are coded and validated, possibly using automated tools and the design engine, the module simulations are available in the VP library. To simulate a system, the VP engine requires a design characterization, as previously discussed. Suppose that the network consists of two processor modules, a system interface module, and a dual redundant parallel bus, as shown in Figure 10-6. For the input message mix, the VP engine can produce:

- End to end message latency
- Software process scheduling conflicts
- Similar results at the system level for this cluster communicating with other clusters using the system interface.

The VP engine can also feed back test results to the design engine if module counts must increase due to more realistic measurements of performance. This can provide an *optimizing design environment*. This is equivalent to an HOL compiler that produces an initial pass at the output but follows this with cleanups to improve speed and size. Another example relates to the microcircuit metaphor. The synthesis tools that convert the behavioral simulations into a physical microcircuit description often have separate optimization tools to reduce gate counts. However, practical considerations such as complexity and computing power may limit the use of optimization in the

system ⇒	AA	AB	BA	BB
factors ⇊				
Development Cost in $K	2500	2200	2000	1800
Unit Cost for 1,00 in $K	150	130	110	90
Schedule in months	100	90	80	77
Performance Reserves %	50	60	70	80

Figure 10-8   Possible ranking factors for candidates.

computer design arena. As work station performance increases and software tools mature, optimized design will become more important because of its large payoffs in reduced design time.

The virtual trade studies might rank candidate on the basis of performance, cost, and schedule factors as shown in Figure 10-8. A final issue is the output section of the VP engine. The engineering team must present the candidates to the customer and *justify the recommendations*. This can often be quite contentious as more detail is added to such top level summaries. Any weakness in the engineering report rapidly turns smoke into fire. The VP engine can allow a more rapid response to such crises.

10.4.2 Product synthesis

Design synthesis is a powerful capability in microcircuit design. This involves automated tools that can translate the behavioral description of the microcircuit into a physical description. The behavioral description includes functions, interfaces, and performance while the physical description provides gate and signal pin locations and characteristics. The physical description, when properly reformatted, can drive numerically controlled machinery that manufactures the device.

The physical descriptions of the circuit cards and backplane assembly consist of the printed wiring board (PWB) files and module assembly drawings. These are elements of the circuit card design synthesis, not system synthesis. Nevertheless, the automated design capture of these elements represents another analogy. The inputs to the layout tool consist of component lists and a physical description of each component extracted from a large database. The physical description includes the device size and location of each signal line of the microcircuit. An automatic routing tool can generate a three dimensional model that represents the signal traces between microcircuits on a multiple layer PWB. This tool can also generate a Netlist for the

PWB that controls manufacturing equipment. This equipment can produce PWBs at a remarkable rate. Once the design team submits a validated Netlist, the specialty house can often produce the first PWB within 24 hours.

In order to establish requirements for a similar capability in computer system design and manufacturing, one must focus on the outputs that correspond to a physical description of the system. The outputs that lead to manufacturing are primarily a combination of engineering drawings and documents.

Several drawing packages and assembly manuals are the basis of manufacturing efforts. Although the specifics depend on the nature of the product, a few examples may illustrate the potential of automated synthesis.

The wiring harness between the computer circuit boards and the front panel connectors provides an example. An engineering team might devise a simulation capability that measures the distance between each end point, considers installation factors, and produces a description. The installation factors involve routing concerns such as locations of cable runs and attachment points for mechanical stress relief. The description includes a description of each wire in the cable as well as the assembly instructions. The description of a wire might include the wire gauge, color, and length. An assembly drawing could show the relationship of each wire to the others in the cable assembly. An automated machine might manufacture the cable from this description. This might include electrical continuity tests as part of the quality control process.

The front panel provides another example. Even though the computer design might rely on a standard enclosure size, the placement of connectors on the front panel is usually specific to the application. A blueprint of this assembly in electronic form can serve as an output to automated milling and drilling machines that drill holes for connectors and fasteners.

Most of the remaining assembly of the computer involves rather large objects that are easy to manipulate. Therefore, the final assembly steps might be manual. However, the testing of the final product is likely to be highly automated. Such tests are run at the factory as part of the quality assurance process. They involve electrical, safety, and workmanship checks as well as the electrical burn in. The latter test simply consists of allowing the computer to run for many hours and then rerunning the functional tests to eliminate a statistical phenomenon called *infant mortality*.

This chapter deals with the VP outputs that interact with the manufacturing processes for the system. If VP system can automatically produce these outputs, the VP engine is capable of design synthesis.

10.5 Status of automated prototyping

The three stages of product development are, once again, design, prototyping, and manufacturing. Virtual methods provide powerful capabilities for each element. The last chapter examined virtual design methods. This chapter examined virtual prototyping. The topic of virtual manufacturing is outside the scope of this textbook, but represents an active, rich area of both theoretical investigations and engineering applications. Previous discussions described some of the automation methods of virtual manufacturing.

VP presents a broad vista across many problem domains. As a result, large scale enterprises, often driven by cost or schedule desperation, might typically use large scale VP. The other extreme is very simple problems. The implementations are slowed by the cost of software development, computing resources needed, and immaturity of the underlying theory. Progress is likely to solve all of these issues and make VP much more practical.

VP methods apply to both virtual trade studies and product synthesis. The elements of virtual trade studies are emerging from many demonstration projects today. Product synthesis is a newer area for VP. The synthesis capability might be tied directly to the design engine for simple enough applications. This could offer the opportunity to feed results from the product synthesis and virtual trade studies back to the design engine as an optimizing system.

References and notes

1. Yourdon, Edward and Constantine, Larry. *Structured Design* (Prentice Hall, Englewood Cliffs, NJ, 1979), p. 148.
2. Silberhatz, A. et al. *Operating System Concepts* (Addison-Wesley, New York, 1991), pp. 106–121.
3. Based on testing by the author. This number seems to be surprisingly consistent.
4. Silberhatz, A. et al. op. cit., pp. 104 ff.
5. Silberhatz, A. et al. op. cit., pp. 195 ff.
6. Robertazzi, Thomas. *Computer Networks and Systems: Queuing Theory and Performance Evaluation* (Springer Verlag, New York, 1990).

Projects

1. Reconsider the notion of end to end message latency for the simple example described below.
   (a) The message mix consists of 10 messages of 64 bytes each. These messages are scheduled at 20 Hz. The overhead per message for transmission on a 20 MHz synchronous bus is:
   - 6 clock cycles for arbitration
   - 4 clock cycles for control information
   - 4 clock cycles for acknowledgment.

   What is the total amount of time necessary for the bus operations? Assume that there are 64 data words per message.
   (b) Suppose that a different software process produces each message. Also, suppose that the burst rate of the 10 processes are {2,4,6,8,10,10,8,6,4,2} time ticks respectively. To "kick off" I/O one must not only update the data block, but also send control information to the I/O interface logic, such as an interrupt or a write to a dedicated memory location. If each process kicks off its own

I/O, what is the scheduling delay for each message? Use the shortest job first algorithm. What is a suitable hard deadline for preemption?

(c) Consider the same scenario for the software processes in item b. What is the scheduling latency for each message if the function of process 10 is to perform all the I/O for the other processes?

(d) Well designed software for an I/O interface might contain several layers of increasingly more abstract logic. The lowest layer must reflect the specifics of the I/O controller operation, but the uppermost layer might provide very general operations, such as PUT for writes or GET for reads. Suppose that three such layers exist and that a procedures call requires 150 µs. Suppose also that procedures exist to set up the control words of this example. What is the total software overhead per message? Quantify other sources of "software overhead," such as responses to an interrupt on completion.

(e) Consider similar calculations for the host side delays on the target node. When a message arrives, it triggers an interrupt that the destination node processor must service in software. This processor must then read the inputs from I/O memory and copy them to program memory. After this, the processor must release the I/O memory for new messages by signaling the I/O controller with an interrupt. Estimate the host side latency on the destination node.

(f) Sum the previous estimates. What is the end to end latency for each message? What percentages of these latencies do bus hardware operations represent? What percentages do software overhead represent?

Significance:

These calculations should demonstrate that software overhead dominates. The effect of this phenomena is that hardware operations are so fast as to be "invisible" to application software. This allows the application programmer to treat communication channels as memory devices and greatly simplifies design. If these characteristics are not met because of poor hardware performance, finely tuned application software might be necessary to assure latencies within acceptable bounds. Software designs of this type are often exceptionally non-portable.

2. Obtain a textbook on operating systems, such as Ref. 2.

(a) What functions do an operating system provide for an application process?

(b) Describe the software process scheduling algorithms. What are the advantages and disadvantages of each?

(c) What types of systems are suitable for on demand scheduling? Give some examples. Cite some examples of applications that are not suitable for on demand scheduling. What are the properties of such systems that make the on demand approach unsuitable?

(d) What are some alternatives to on demand scheduling? Which of these examples provides the most consistent, predictable execution sequences? In what types of systems is predictability important? Can such systems also respond to asynchronous events? If so, how?

3. Consider the output of the VP engine.

(a) What are some candidate ranking factors for a new design? Include the factors of this chapter, but provide more detail. What other factors are applicable? Consider this from the perspective of a customer who has a certain budget and schedule.

(b) What are some candidate ranking factors for a reengineering project as described in Chapter 8? What overlaps are there with the factors for a new design? What factors differ?

(c) Consider that the VP engine is making a recommendation to you for a personal computer purchase. What factors would you like the VP engine to emphasize? If you have not done so recently, visit an electronics store and ask the same questions of a sales person. What was the emphasis of the presentation by the sales person? What factors did the sales person emphasize that annoyed or did not convince you?

(d) Describe the systems that your customers use. What are the factors that might convince your customers to make a design change?

4. Consider the inputs to the VP engine.

(a) Identify the elements, or "building blocks," of systems that you use or design. What are the key features of these elements that should be fed to a VP engine?

(b) How do these elements relate to the candidate ranking factors? For example, if performance is a secondary consideration subordinate to cost, how might you characterize the inputs to reflect these factors?

(c) Discuss how the systems of question 4(a) are amenable to electronic descriptions. How might such systems appear on a STD? What might be the features of a VHDL behavioral simulation of such a system?

(d) Discuss the behavior of an expert system to perform these inputs. What tools described in Chapter 5 might produce outputs that feed naturally into the VP engine?

5. Consider the design rules of the VP engine.

(a) What types of simulation models might be found in the library for a computer design VP engine?

(b) What initialization or configuration might such models require to operate together as a system simulation?

(c) What type of computer might host such simulations? Consider the availability of tools and the performance of the tools on the host system.

(d) Discuss some features of the test scenario inputs. For the performance factors, what are some considerations in running the test? What features must you measure and report from the VP engine to reflect the test conditions?

Part III

Synergism of OOD and VP

11 Design trends and issues   **223**

Previous discussions lay the foundation for a strategic vision of computer design automation. Successful execution of a strategy requires timing, so it is important to understand why this discernment can transcend the realm of visionaries to that of practitioners. As a model of this maturation, this chapter returns to a broad scope that encompasses the processes, technologies, and methods. The discussion includes a statement of the significance of each and the trends in these areas.

chapter eleven

Design trends and issues

11.1 Systems engineering standards

The various systems engineering standards that this text describes are important to VP for several reasons. The standards identify the engineering and management practices associated with various phases of a program. This includes the inputs and outputs from each phase. The standards also define the requirements and partitioning processes that lead to validated technical descriptions of the product. Finally, such standards coupled with information about automated tools can suggest appropriate engineering methods. This approach may lead to OOD or some other design method.

Although methods often dominate the attention of a development team, process has an equally strong impact. Since the process models represent an abstraction of the engineeering and management activities, they represent a template for such tasks across many projects. For a fully integrated process, it is important to have compatible process models for both software and system design. The process then forms the environment in which methods that emphasize model building can flourish. One such method that engineers can apply successfully to both software and systems engineering is OOD. Engineering standards formalize process models and provide suitability guidelines for methods.

Previous chapters describe some work in formal systems engineering standards and process models. This work is ongoing and has at least one deficiency in its current incarnation. The military systems engineering standard, MIL-STD-499B, describes all of the usual risk reduction and management activities. However, the standard explicitly builds its edifice upon the military and aerospace acquisition system and certification procedures by reference to other standards. The difficulty in applying this standard to large commercial projects is to find suitable replacements for such elements.

The current commercial systems engineering standard is IEEE-1220. This standard also implies the existence of an acquisition system and certification procedures, but does not identify them. In the military system, these procedures are defined in law to build audit trails for contracts officials. It is unclear whether the commercial market will drive the development of comparable standards. The impact is that military users of MIL-

STD-499B are unlikely to have difficulty in complying with IEEE-1220, but commercial users may struggle to apply the standard to particular problem domains.

Standards unique to an agency or project may also emerge periodically. Some agencies may formulate a process for each project or with additional, formal standards. For example, suppose that the Federal Aviation Administration (FAA) wishes to consider the certification of a complex, highly integrated, new avionic system. The customer reviews in this case are with engineers from the airlines, but certification is the responsibility of FAA engineers and managers. This is quite different from the military system in which the customer and the certifying agency are the same entity. Similar paradoxes may emerge for the design and certification of space electronics by the National Aeronautics and Space Agency, nuclear reactors for the Department of Energy, ground based communications networks for FAA, and a host of others. The result may be that each agency must formulate its own acquisition and certification models. Whether this renders system engineering process standards inapplicable is a currently evolving answer.

The complexity of the design must also be a factor in tailoring an engineering standard. For military systems, the complexity of the design is formalized in a series of acquisition categories. The most complex systems require the most formal reviews, audits, and contracting procedures. The acquisition of simpler, relatively low cost systems involves correspondingly less arduous management.

The same approaches seem appealing for commercial computer equipment, but the limits of the thresholds and categories are unclear. The vendor of a single replacement hardware card in a personal computer should not require the costly, time consuming scrutiny that a new design for a computer network deserves. The problem centers on the tailoring standards for applications. How can a customer demonstrate when a design review is beneficial or simply a waste of time and money?

Military standards such as MIL-STD-499B require or imply numerous related procedures. IEEE-1220 is an important step forward in bringing the formal design methods honed for military systems to commercial systems. However, its goal of a universal process model may be unattainable. This is a difficult problem, but is a crucial element of design automation because it defines management risk approaches and deliverables. In the OOD terminology, process models are likely to represent a class with many customer object instantiations. What, then, are the identifying traits of this class?

Just as OOD can allow abstract model building of a design problem, it can also serve as a tool to generalize the MIL-STD-499B model. Figure 11-1 is a beginning in defining such traits. In concordance with modern management practice, as well as common sense, all actions and reactions must stem from customer needs. Usually, the customer presents computer system needs in operational terms, not as design requirements. Therefore, a requirements formulation stage is the entry point into the design process. However, the customer and design engineers must validate these requirements to prove that the formulation is correct. The iterative, formal review cycle demon-

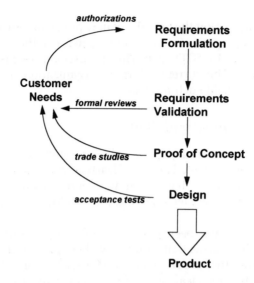

Figure 11-1 Systems engineering process traits.

strates agreement between the customer and designers that the requirements capture is correct. At some point of requirements maturity, prototyping begins to serve as a proof of concept. The results of trade studies validate the requirements capture and feed back into the customer review cycle. As a baseline design emerges, the customer and design team must agree upon acceptance tests. Early in the development, this may simply be an agreement that each requirement is testable. Later in the process, reviewers must agree on tests for each requirement. After successful completion of acceptance tests, the design can enter the production phase. The five classes of this model are:

- Customer needs
- Requirements formulation
- Requirements validation
- Proof of concept
- Design.

If one treats "customer needs" as a class in the model, there could be objects for the various customers or problem domains. The operations on this class might include formulation of a need statement and comparison of requirements to this statement. In the latter case, the output is an enumeration type, {approved, disapproved}. The remaining elements might also serve as classes. Requirements formulation consists of technically defining the system, often using other systems as templates. Requirements validation involves comparing these requirements to historical data and estimates of technology capability. The proof of concept is the true test of the technology and integration approach. The product baseline emerges from the design stage.

In order to fully exploit the benefits of new policy initiatives such as dual technology use, leading practitioners must generalize the system engineering process model of MIL-STD-499B. This discussion presents an example of such generalization. The current efforts of organizations such as NCOSE, IEEE, and SAE will yield such models.

11.2 Expert systems revisited

An expert system is a simulation of a dialog between an expert in some technical area and an interested novice. Clearly, some elements of the process model are more conducive than others to expert systems technology. This discussion examines these elements and the specific technology underpinnings.

Unfortunately, AI has been the source of numerous disappointments. Developers of early systems that could play chess promised "thinking" and "learning" machines that proved elusive. More delays and discouragement emerged when it became clear that sufficient computer performance was a major impediment to implementation of expert systems. Early systems often relied on brute force search and parse algorithms that demanded vast storage capacity and execution speed. More recently, research in neural networks was advertised as emulating the biological processes of the human brain and therefore its capabilities. These claims have also been largely discredited. However, current theories coupled with high performance work stations place users on the threshold of important new computing capabilities. The breakthroughs are occurring nearly simultaneously in several areas.

The first area involves those design aspects that a modeler can reduce to algorithms, such as those of Chapters 8 and 9. These algorithms are critical because the capture of the design rules in a database is a precursor to expert systems integration. The control logic that selects applicable rules must also be tractable. The practical experience that engineering teams garner with this rapidly maturing technology provides experience, requirements, examples, and software components that new projects might reuse.

Another historical problem is the testability of the inferences (rules). Currently, this may involve comparison of predicted results to hardware measurements and historical data. This is often a laborious task whose magnitude may rival that of the development. Ultimately, though, this testing process can also be mechanized to decrease cycle time. As with other software applications, automated testing can offer important productivity gains.

Theorists have largely completed the formal elements of natural language and synthetic vision. However, the high computer performance needs for such systems have slowed implementation. Current generation work stations provide the threshold capabilities for such implementation, so this technology is likely to emerge into mainstream usage in the next five years.

The final concern is how to blend the human operator into the system seamlessly. Clearly, it is easy for a high performance work station to overwhelm a human with dazzling graphics, full range sound, and challenging tasks that require inputs. The classic machine intelligence test goal can be a guide to the VP system developer. Imagine that the computer and operator are located in different rooms and must communicate through viewing screens and audio devices. The goal, stated in simplest terms, is to make it impossible for the human operator to discern whether a human expert or the expert computer system is on the other side of the wall. This is challenging for many reasons, including the element of personality and presentation style that many human experts use effectively to solidify key recommendations. The distinction between data and information is not always easy to quantify. However, such experts often carefully limit extraneous material to make the presentation convincing and clear.

The appropriate strategy for implementing expert systems may be incremental. The design engine might be the first application since, at least in the initial stages, trained humans could operate the system and interpret its results. However, the rules and control logic involved in such design can be subtle, so experience with such systems is important. The requirements arena is probably a more tractable short-term application. Many automated tools now exist that provide a basic capability of this type. However, it is difficult to imagine that managers could justify the expense of expert systems simply for requirements management because of usage statistics vs. cost. This technology must be coupled to some other automated design tool to make its use economical. The VP engine presents design outputs and recommendations to experts for presentation to customers.

The final area is the customer interface. An expert system might speed initial captures of requirements information. In the short term, though, it is unlikely that computers will sell computers even if computers design computers. The human interaction may be simply too complex to simulate. However, this process offers an opportunity to demonstrate the capabilities of such systems to new customers, so a blend of human and automated customer interfaces will likely be used.

The strategy for an implementation might follow this pattern:

- Phase 1: extensive use of expert systems design coupled to current generation requirements tools
- Phase 2: coupling of manufacturing software interfaces to the design engine outputs
- Phase 3: fully automated design and prototyping
- Phase 4: some automated requirements capture from customers.

The schedule for the first three phases might involve a 10 year span. The first phase might require 5 years to fully implement and test, the third phase could follow rather quickly in about 2 years, and the final phase might require an additional 3 years. There could also be some overlap in this schedule with sufficient resources, so around 7 years is probably a more realistic target. The final phase probably requires more analysis and research.

Expert systems unquestionably are an important element of rapid prototyping. The goal of such systems must be to help humans be more creative by relieving them of the burdens of repetitive calculations and testing.

11.3 The significance of OOD

A general trend in system design involves highly integrated solutions. Functional integration lies at the core of this approach. This involves massing closely related system functions at the same processing node instead of providing one node per funtion. The latter approach was often driven by limitations in key technologies. However, the former approach can quickly result in exceptionally complex interconnect schemes.

The subtlety of programming such hardware demands that software requirements analysis be performed in parallel with similar tasks for the hardware design. It is simply too complex to add embedded software to such systems after hardware production begins. The historical trends in software complexity management offer intriguing parallels to similar problems that designers of highly integrated architectures face.

As computing hardware became more capable in the 1980s, software engineers delivered products of unprecedented complexity. Unfortunately, formal methods of complexity management were unrefined and not widely used. In addition, many large programs attempted to treat software in much the same manner as hardware black boxes in which isolated design teams performed requirements analysis. This led to some dramatic failures of managers to predict delivery cost and schedule, which culminated in the celebrated *software crisis* proclaimed by several government customers. In order to fully appreciate the significance of OOD, it is critical to replay this drama.

Despite the zealotry of purists, the theoretical foundations of both SD and OOD stem from such concerns. However, the focus of each method is slightly different. SA/SD relies on sequential development phases that have distinct starting and ending points. The driving forces of OOA/OOD are the iterative nature of development and an increasing importance of reusability.

The only practical method of developing an extremely complex system is to partition the top level design entity into smaller elements that are individually tractable. However, as the number of team members increases, the likelihood of problems with integration increases. Therefore, the managers of such a development must account for an evolving design baseline instead of a small number of stepwise captures. The cleverness of OOA/OOD lies in the recognition by its formulators and proponents that solutions for managing such an iterative process also offer powerful reuse methods. This relationship stems from the nature of tasks for both new and reused designs. As with the development of complex software applications, requirements analysis and product testing tend to be the most costly portions of the engineering efforts. Code or data entry tends to represent a small fraction of development cost, although this is often counterintuitive.

This observation illustrates two important policy and management goals that relate directly to risk reduction. The first is *design for testability* in which one aspect of requirements validation is early proof that a certification test is

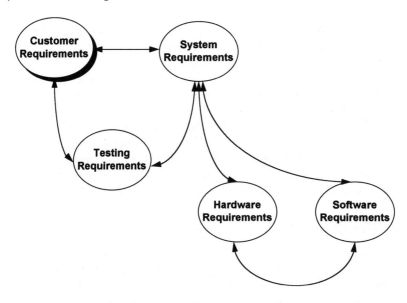

Figure 11-2 Iterative design process.

definable. The second aspect is *rapid prototyping* in which engineers apply the tests to models of evolving design baselines. The capability to iterate requirements and product baselines quickly is a tenet of OOA/OOD. VP provides the rapid prototyping system. Therefore, the OOD and VP play complementary roles in an overall risk management strategy.

Figure 11-2 illustrates the iterative nature of design. Engineering teams must anticipate that customer requirements are as likely to change as detailed requirements. This may occur due to redirection on cost or schedule, a refinement of the system's operational concept, or perceptions of technical difficulty. The customer interacts primarily with engineers involved in acceptance testing and top level system requirements. The hardware and software lower tier requirements flow from the upper level requirements. Therefore, changes in any of the top tier requirements tend to ripple through the entire design process. Modern management approaches stress the need to adapt to constant change. The engineering teams must learn this lesson also.

Another important trend is the role that software plays in complex systems. In many respects, hardware elements are easier to integrate than ever due to the emergence of mature interface, packaging, and electrical standards. The expertise of early system design engineers was often targeted at understanding the subtle differences among vendor implementations for such hardware. This is still important, but the competition and standards have driven out many poorly executed first generation designs. For these reasons, software design must represent an increasingly large percentage of the experience base for system integrators. In order to simplify the design process, system designers and software engineers must use identical methods to describe the product baseline. Any translation between the two design spheres injects risk of garbled requirements definition and delays in response. Therefore, system engineers are more important than ever in the successful design of such systems, but common methods are essential to assure that the previous generation crises are not repeated.

Practicality is another concern. If common design approaches for hardware and software are beneficial, then OOA/OOD should be the approach of choice. The trend is in favor of OOD in terms of research, education, and tool development. University credit and continuing engineering education classes heavily emphasize OOD. In addition, automated engineering tool vendors are heavily marketing products for this type of design. The impact of this trend is that the next generation of software engineers will be as comfortable with OOA/OOD as current engineers are with SA/SD.

11.4 Trends in enabling technologies

Three general elements provide a firm technology basis for VP systems. The first area is computer hardware, especially key factors that drive software performance. Such factors are processor execution speed, memory density, and storage capacity. The second element is software technology, which naturally lags somewhat behind hardware capability. However, the personal computer and work station technologies are allowing software an unrivaled growth era. The third element is specialized I/O interfaces and networks.

The single technology area that leads others as an enabling capability for VP is the constantly increasing computing power of work stations. New processor designs have increased the execution speed of such systems tenfold in 7 years. Disk capacity doubles approximately every 4 years, while memory storage capacity doubles approximately every 2 years. Graphics accelerator hardware and display technology produce images that rival photography. These trends began in the middle 1980s and are likely to continue until the end of this decade.

Figure 11-3 illustrates these trends by showing the projected growth based on a unit value for each for the starting time. These trends also have the impact of driving down work station cost. This is a powerful technology bootstrapping force because this trend makes the host computers more accessible to small businesses that might develop new software or hardware upgrades. It also makes the technology more available to university students for educational and research applications.

The 1980s may have been the decade for hardware designers, but software engineering is likely to fully blossom in the 1990s. One driving force is the hardware trends just presented. However, new developments in design methods, languages, operating systems, and simulation technology also provide building blocks for important breakthroughs. One such element is expert systems. Ten years ago such systems required specialized and expensive hardware and software to assure adequate response time. The software is now available for many popular work stations in C++, Ada, Pascal, or other common languages. Graphics software represents another growth area. New algorithms, such as fractals, hidden line removal, color blending, and other elements provide more capability and performance.

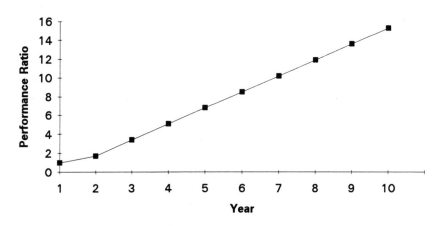

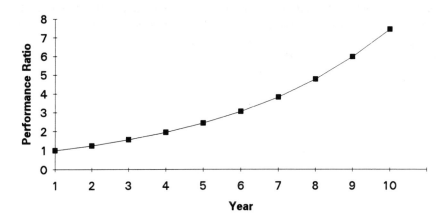

Figure 11-3   Technology trends.

The final leg of the VP triad is the panoply of I/O devices. These include speech synthesizers, high performance stereoscopic graphics, and position sensors. Once again, hardware capability trends in this area are growing, but competition is driving costs down and increasing availability. A related issue is access to computer networks, which allows a wider exchange of information. This includes distributed processing simulations involving many on many engagements.

VP unquestionably requires high performance hardware and complex software. However, the technology trends make available systems priced for the masses that were only research interests 20 years prior. A projection of this trend even a decade forward generates a breathtaking vista. In the efforts to reduce manufacturing time and cost, it is difficult to ignore such capabili-

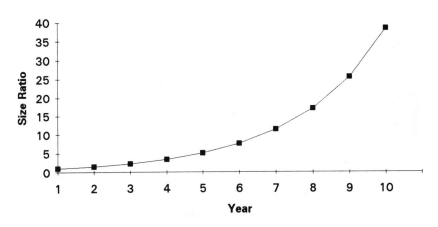

Figure 11-3 (continued)   Technology trends.

ties. Rapid prototyping and virtual reality will ride the coattails of such trends, driven by client demands to increase the quality of the products that they manufacture. For these reasons, VP will emerge from the laboratories and provide a new design era. This process may bootstrap as designers apply VP to automated computer design in order to improve performance and quality, as well as reduce cycle time.

Project approaches and solutions

Chapter one

There are no exercises for the first chapter.

Chapter two

1(a). MIL-STD-499B, despite two years of effort and much labor expended, this standard has fallen to in the rush to acquisition reform. Currently, the National Council on Systems Engineering (NCOSE) has accepted responsibility for the latest draft of MIL-STD-499B. This suggests that a discussion of MIL-STD-499B is still relevant. In addition, some engineering firms already use this standard despite the fact that it has not been approved for Department of Defense acquisition.

The most mature document in comparison to MIL-STD-499B is IEEE P1220. The IEEE document is much more general and it is likely that any program that complies with MIL-STD-499B will also comply with IEEE 1220. The primary distinction between the two is that MIL-STD-499B identifies an acquisition process by reference, but the IEEE standard leaves this undefined. The selection of the acquisition method obviously has a major impact on the system engineering process. Similar comments apply to formal reviews.

Finally, the SAE has several systems engineering methods. The most relevant is a task group within AS-1 that is developing a systems engineering handbook.

1(b). The following MIL-STD-499B definitions apply only to process: life cycle, user, systems engineering process.

1(c). The primary process related block is called "System Analysis and Control."

2. The solution to this problem depends upon the tool vendors polled. Point out in the literature any obvious marketing hyperbole. The tool vendors know that object oriented technology sells. Some vendors make it a point to emphasize that a tool can be used with either method.

3(a), (b). The CMM addresses a different problem than MIL-STD-499B. Although the two have some overlaps, there is no direct conflict. The CMM describes an iterative, optimizing method for improving an organization. MIL-STD-499B addresses the relationship among engineering and management tasks. The relationships are the same whether a particular organization accomplishes them efficiently or otherwise.

3(c). Once again, CMM is not a method for risk assessment. Refer to the discussion of Technical Performance Measures in MIL-STD-499B. However, there can be some confusion because of the use of "process metrics" in CMM implementations. The process metrics measure how efficient an organization is, not necessarily how fast the quality of a product is improving.

A discussion of the integrated product team (IPT) approach might be appropriate. The IPT structures the organization around the end item product. This has the effect of coupling process improvement to product improvement.

3(d). CMM supports either management approach. Perhaps the most popular approach is a matrix solution to the IPT. Often, the IPT consists of area leaders for each product subarea and engineering subarea. The table below shows a simple example for the circuit cards of an embedded computer. Each "X" represents a team leader (known in former times as a deputy program manager). The categories along the top represent "line management" whose function centers on providing trained engineers for each subarea. The rows represent the project leaders for each engineering subarea associated with the final product.

Software	Systems engineering	Reliability	Logistics	
X	X	X	X	Data processor
X	X	X	X	Graphics processor
X	X	X	X	Input/output

4(a). Many of the risk areas are independent of the specific technology. These might include product maturity, component (or software tool) cost, availability of multiple suppliers, performance, or functionality.

4(b), (c). A risk matrix for a data processor might be similar to the one below.

	DEVICE A	DEVICE B
MATURITY (20%)	high	low
COST (20%)	medium	medium
NUMBER OF SUPPLIERS > 2 (20%)	yes	no
PERFORMANCE (20%)	medium	high
FUNCTIONALITY (20%)	yes	yes
SCORE	120	80

Key: (high, medium, low) = (2,1,0); (yes, no) = (1,0); 220 points possible

This example illustrates how numbers can average out important features. High performance should probably not be the sole criterion, but functionality might be a "show stopper."

Of the 220 points possible, a total of 154 represents 70% and 110 represents 50%. These might be the cut-offs for rankings of high, medium, and

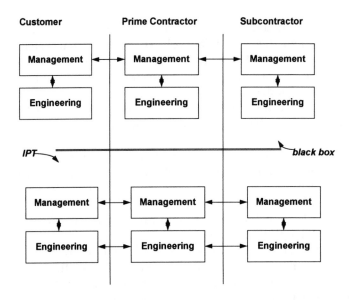

Diagram 1 Organizational entities used in Chapter 2, Project 5.

low. Therefore, candidate A above is medium risk while candidate B is high risk.

5(a). The "black box" organization refers to management of individual commodity areas such as different computer types. The chief defect of this approach is its insularity. It is difficult for engineers in a large organization to share lessons learned. In the "black box" approach, a similar organizational structure replicates within each level of the overall structure from customer, prime contractor, to subcontractor. In principle, the IPT approach involves a horizontal strategy in which engineers from the customer, prime, and suborganizations each work together without communicating through individual management layers.

5(b). Diagram 1 illustrates this distinction in more detail. In the IPT method (bottom of the diagram), engineering is "concurrent." In the black box method, engineering proceeds in a more serial fashion. The advantage of IPT is efficiency, but its disadvantages may include dilution of responsibility and authority. For these reasons, many key individuals can feel quite threatened when transitioning from a black box organization to an IPT.

Chapter three

1. The solution depends upon the system that the student selects. There are examples in the text, but the purpose is to provide some practice in decomposing a system into classes and objects. Refer to Booch for examples also.

2. The purpose of this exercise is to perform the classical systems engineering tasks called "functional partitioning" and "functional decomposition." MIL-STD-499B provides some formal definitions, but an intuitive approach may be more appropriate at this stage. The class diagram of the

previous exercise should lead the student to pattern the decomposition after the class definition. This is the relationship between formal system engineering methods and OOD. However, this need not be true as an SA/SD approach could lead to a different decomposition. Chapter 4 addresses these distinctions in more detail.

3, 4. To construct these diagrams, refer to Yourdon (Ref. 4, Chapter 3) for examples and rules. Chapter 4 also provides examples and a summary of rules.

5. The purpose of this exercise is to demonstrate the need to fully understand customer requirements, not simply respond to jargon. Often, customers do not precisely understand how to solve a problem (if they did, they would not need your help!), so a clear picture of the project goal is essential. However, the development team should propose the solution, not the customer. Some examples may help.

Employment ads for "systems engineers" often represent a spectrum of job functions. In the arena of information systems, the title often applies to the individuals responsible for local area networks. The title can also refer to engineering management or formal methods, such as reviews, specifications, and standards. Finally, the title can apply to engineers responsible for the integration of hardware and software. Imagine how much confusion could be caused if the customer simply stated a need for engineering support as "I need systems engineers."

The situation may be similar with regard to software engineers. There is a spectrum of development environments, one of which may be implicit in the customer request. "Object oriented design" often means that the customer has recently purchased an expensive tool. The development team must understand that selection of a particular tool may lock out certain design approaches. Usually, it is important to resolve such misunderstandings early in the development.

Chapter four

1. The requirements matrix might resemble the table below.

Operational requirement	Functional	Performance	Interfaces
O1: Start-up	Perform alignment	< 5 minutes	Angular sensor external position [see below]
O2: Navigation	Perform navigation	< 1 Nautical mile/ hour error	[See below]
O3: Sensor capability	Measure angles and rates	< 3g's; < 10 radiens/s	None
O4: Waypoints	Operator entry	[See below]	[See below]

The entries (see below) indicate derived requirements and might appear as follows:

O1, interfaces: Global Positioning System might be a good candidate for this operation. Other methods include a transfer of alignment (latitude, longitude, altitude) from some other source such as another aircraft or a ground station.

O2, interfaces   Over long periods of time, this accuracy requirement will require inputs from other sources. Actual implementations typically use barometric altitude, Doppler velocity sensors, and Global Positioning System.

O4, performance   The requirement should state something about how much the navigation solution can be corrected on the basis of periodic inputs from the operator. For example, if the solution degrades to 2 nautical miles per hour, the operator might use several "fly over" waypoints to bring the solutions back within error bounds.

O4, interfaces   Some type of display and keyboard interface is necessary to implement this function.

2. Number the derived requirements listed above as R1–R4. The test matrix might appear as below:

R1   Test 1: Perform navigation simulation for a typical mission scenario to demonstrate proper initialization.
        Test 2: Execute initialization code on a prototype using simulated external alignments.
        Test 3: Execute initialization code on a fully functional prototype.

R2   Test 1: Perform navigation simulation for a typical mission scenario to demonstrate proper performance with and without external aiding.
        Test 2: Execute navigation code on a prototype using simulated external navigation aids (such as Global Positionig System).
        Test 3: Execute navigation code on a fully functional prototype.

R3   Test 1: Perform navigation simulation for a typical mission scenario to demonstrate proper performance with and without operator inputs.
        Test 2: Execute navigation code on a prototype using simulated operator inputs.
        Test 3: Execute navigation code on a fully functional prototype.

R4   Test 1: This might involve some type of human factors criteria to simplify entry. This could involve display formats, keyboard size and shape, sunlight readablity, and similar criteria.

3. The format of this summary could either be a table or a graph. However, it should illustrate the relationships among the tests. For example, R3 Test 3 and R4 Test 1 could be run simultaneously for maximum efficiency. This is allowable because the tests are independent. However, R2 Test 3 and R3 Test 3 could not be run simultaneously because each provides alignment aids. These aids must be tested separately because the requirements state performance for each type of input, not for both types.

4. The STD might appear similar to the one shown in Diagram 2. The state definitions should include an initialization sequence, error conditions, and normal operating conditions. The student should be prepared to defend a state definition based on the system behavior and the interfaces used for each. An example of this is given in the following table.

State	Behavior	Interfaces
Initializing	Start-up sequence	External position fix, such as Global Positioning System, internal sensor system
Navigating	Provides position and orientation solutions	Internal sensor system
Displaying	Outputs the navigation solutions to a user display	Display hardware
Updating	Provides corrections to the solutions	External radio navigation, such as Global Positioning System. Also, operator waypoints entry.
Position error	Indicates an unsuccessful attempt at update (see below)	The operator should be able to override this and continue navigating.
Terminate	The system has suffered an unrecoverable error; this is a system failure	Not applicable.

Note: The position error state occurs when the difference between the calculated position and the update entry is large enough as to be unresolvable. This resolution usually depends upon the type of update.

5. A timing diagram for the operator updates might resemble Diagram 3. The key parameter is the response time which defines how quickly the operator can enter waypoints. If this response is not fast enough, then the data entry system would have to buffer these locally. If too much time passes, then the navigation solutions may not be correctable. Timing diagrams for other update mechanisms might also be useful to construct (see next question). Finally, simple timing diagrams for initialization (start-up) and terminate (shutdown) could be useful to discuss. The operational details of the navigating state are probably beyond the scope of this presentation, though.

6. The class of sensors might include the following list of objects.

accelerometers: Internal to the navigation system, provides linear acceleration. Provides a measured output using a dedicated, high speed interface. Three needed:

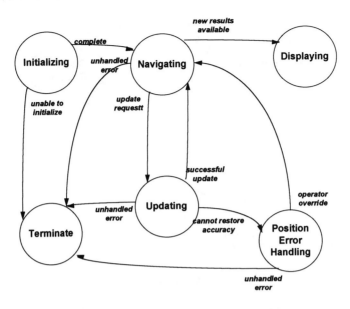

Diagram 2   Chapter 4, question 4 STD.

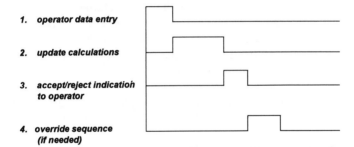

1. *operator data entry*

2. *update calculations*

3. *accept/reject indicatioń to operator*

4. *override sequence (if needed)*

Diagram 3   Chapter 4, question 5.

accelerometers.x_body
accelerometers.y_body
accelerometers.z_body

angular rates: Internal to the navigation system, provides angular velocities. Provides a measured output using a dedicated, high speed interface. Three needed:
angular_rates.x_rotation
angular_rates.y_rotation
angular_rates.z_rotation.

position updates: These might include the various position aids such as operator inputs, Global Positioning System (GPS), or Doppler Velocity Sensor (DVS).
updates.operator
updates.GPS
updates.DVS

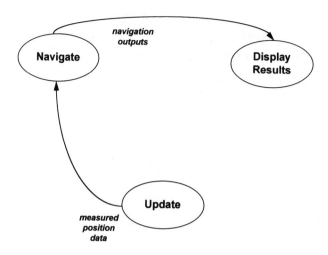

Diagram 4   Chapter 4, question 7.

The navigation and update code could also be broken down into classes and objects, but this discussion is probably beyond the scope of this problem.

7. Diagram 4 shows a top level DFD that includes the three processes used in normal operation. It probably is unreasonable to expect more detail from the students unless they have a navigation background.

8. One must select a perspective from which to view the classes. According to Rumbaugh (Ref. 7 of Chapter 4):

> *A module is a set of classes that capture some logical*
> *subset of the entire model.*

The subsetting strategy might be based in this example on the top level system states. Define a superstate called "normal operation" that consists of the states navigating, updating, and displaying. The module diagram might appear similar to the one shown in Diagram 5.

You may wish to focus on the more generic aspects of the diagram, such as the interaction between the displayed solutions that the system presents to the navigator and the keyboard entries that this operator provides as waypoints.

Chapter five

1(a). Numerous design vendors offer such literature. One example is LSI Logic.

1(b). The behavioral models should include primitive "cell" functions. These might include logic units such as adders, multipliers, or shifters.

1(c). Many of the tool vendors allow users to build up their own custom libraries. These might include macrocell functions that the design team uses frequently but which are not available in the library that the tool vendor

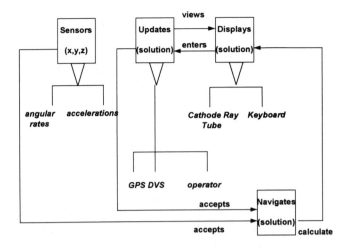

Diagram 5 Chapter 4, question 8.

provides. If the design of these custom functions is based on modern software methods, they can be made reusable to many projects.

1(d). Superficially, microcircuit design resembles the process of Figure 5-7. The requirements are the basis of a product description that captures the behavior, control, and interface processing of the design. However, the metaphor begins to unravel as one attempts to formulate the physical description of the design. In microcircuit design, a synthesis capability allows a team to move from the behavioral description to a physical design using highly automated methods. No similar, general capability exists in the systems engineering realm.

2(a). The CMM captures the iterative nature of design, but it models the engineering management process, not actual engineering tasks. Figure 5-7 is an attempt to augment the description of the CMM in this regard.

2(b). Many Ada and C++ vendors offer an ISA simulator that allows a software developer to execute code. This can be quite useful if hardware is unavailable or is so new that it still does not function properly. However, ISA simulators usually only model the processor internals and memory interface. It may not, for example, simulate interrupts or I/O activity. Therefore, such systems are only of limited value in developing real-time code for which timing and the execution environment are critical to proper operation.

2(c). See the discussion under 2(b). To add these features, one simply needs to add software packages to simulate these functions. Unfortunately, they are often very implementation specific, so it may not be cost effective for an Ada vendor to provide such tools. If hardware is not available, the user might need to work with the Ada vendor to integrate custom I/O and interrupts packages.

3(a). Analysis of test coverage usually involves building a diagram (or simulation) that connects all possible combinations of inputs and outputs. The test conceptually involves setting the bits for each combination of inputs and then comparing the observed output bit patterns to values stored in a test table.

A single bit adder provides a simple example. If each input is equally probable, then the controls and inputs are related as shown in the table below.

Input 1	Input 2	Output 1	Output 2
0	0	0	0
1	0	1	0
0	1	1	0
1	1	0	1

Each combination is equally likely, so the test should include all combinations. If the test excludes one of the rows (combinations), then it provides only 75% coverage. Elimination of two rows offers only 50% coverage.

In complex design, there might be thousands of combinations of inputs and outputs. Usually, it is necessary to prioritize these to make the on-line test more tractable. Some combinations might involve very low probabilities and the test designer could safely exclude them from the test patterns.

3(b). The generation of test patterns lends itself to simulation quite readily because of its highly repetitive nature. However, some test teams may feel that the design is so complex that they may have missed some combinations of inputs. Sometimes, random patterns are used to augment these test tables. A random pattern is equivalent to drawing numbers out of two different hats. Each hat has papers with the numbers one up to the number of inputs. The first draw shows how many bits to set. The pattern generator then draws from the second hat the number of times indicated by the first draw. The draw from the second hat indicates which bits to set. Sometimes, the order in setting the bits is also important, so the draw order from the second hat must also be saved.

3(c). The hardware equivalent of this is to attach a random pattern generator to the system input stage. The outputs from the design can be saved as a computer file. If one wishes to use the hardware outputs, then one must accept the notion of a "golden standard." This means that the hardware used to generate the patterns must be assumed to function perfectly.

Upgrade efforts sometimes hinge on the capability to use existing test code and procedures. These requirements are a common instance of the need for a "golden standard" approach just sketched.

3(d). The purpose of this exercise is to initiate thinking about process model building. Chapter 6 revisits this topic extensively. Refer, for example, to Section 6.2 and Figures 6-6 and 6-7 for more details.

Procedures involved in running a test might include:

- Scenario generation, especially message mixes or test vectors
- Equipment installation and initial checkout
- Operator control and execution of tests
- Off-line data analysis
- Scenario or event modification, if necessary.

The figures just mentioned demonstrate some approaches to modeling and automating this process.

4(a). The discussion from 3(d) provides a list of elements. Control involves the timing and commands from the operator and scenario controller. Inputs are the test vectors. Outputs may include "raw data" outputs from the unit under testing or "flagged" data that may indicate a discrepancy. In this case, the unit under testing is the system to which one applies the CIPO model, so it performs the processing.

4(b). Once again, this question may start some discussion of automated testing for complex computer systems. Section 6.2 revisits this topic.

4(c). Automated methods can be highly efficient for normal test procedures. The operator is most efficient in dealing with unpredicted or unusual test conditions and results. The reference to question 3(a) may reinforce the theme of automated test.

5(a). Usually, interfaces are easiest to test for several reasons. First, the message scenarios are often available electronically which allows the test team to easily develop (or use) automated generation of interface code. In addition, the functionality of these tests is rather similar among all computer systems since it involves generating the test inputs and verifying correct outputs.

5(b). Either performance or functional testing may be most difficult depending on the application. In principle, performance testing may be somewhat easier if one can relate the requirements to standard benchmarks for processor throughput, I/O throughput, and code size. Functional testing may require the developers to build up a large library of test software in order to provide a sufficiently broad basis for general usage.

5(c). The VP approach may be used in requirements analysis as well as for product testing. The former role usually involves extensive simulation with hardware testing reserved for high risk areas whose features might be too new or complex to characterize simply. In the latter role, the team should maintain historical data to demonstrate that the predictions from the VP system are valid with respect to hardware tests. This provides a feedback mechanism to improve the reliability and accuracy of the VP system.

Chapter six

1. Students might wish to use reader service cards in trade magazines. Virtual reality is such an important topic that there are many articles from which to choose. However, it is much easier to get on a vendor's mailing list than to escape!

2(a). The answers depend upon the system one selects as the basis of the estimates. Some general trends are eluctable from experience, though. The probability that a prototype of an novel or unusual design should function properly is quite low. Suppose that the probability of success with the first prototype is 0.70, then the probability of success with the second is 0.91. This is still not the 95% solution that some designers use as a threshold, so management may decide upon a third iteration to raise the probability to 0.973.

It is important to realize in this discussion that no errors might actually be found in the first prototype. Instead these iterations may be based on statistical process control or some other quality program. As a result, the iterations refer as much to the test code as the prototype itself. In this regard, simulation offers no advantage. Instead, it is the *time to test* that is the key. It might be quite time consuming to configure, load, and test a computer hardware prototype, but simulations can often run faster than real time with minimal intervention.

2(b). Once again, the solution depends upon the example system. In some cases, the processes are simple enough that hardware prototyping can be inexpensive. On the other hand, virtual prototyping can be quite effective for designs in which the manufacturing process is expensive, time consuming, or occasionally hazardous.

2(c). Some type of production is necessary for hardware prototypes. This need not be organic, but the fabrication can often be subcontracted. Although virtual manufacturing is a rapidly blossoming simulation industry, the advantage of virtual prototyping is that it need only interface to virtual manufacturing, not necessarily encompass it. The same is not true of hardware prototyping. Some manufacturing center must own and operate the equipment to fabricate the hardware. History shows that such facilities are enormously expensive to initiate.

2(d). The "reroll" time is highly application dependent. For embedded computer design, the sequence often involves engineering development models, advanced development models, and pre-production models. The delivery times for these are roughly 4 to 6 months apart. A virtual prototype can often be iterated much faster than this. In microcircuit designs, a cleanup effort might require only a few weeks.

2(e). Imagine a finite element model of a computer used for mechanical vibration testing. Defects are often quickly apparent and can be corrected online with modifications to the mechanical design necessary to damp vibration.

3. Fidelity is one limiting factor. A simulation is an abstract model of a system. Except for some very simple cases, most simulation developers make no pretension about capturing all facets of behavior because the models would run too slowly, be to difficult to test, and expensive to develop.

Another factor is validation. As the complexity of the simulation increases, validation with hardware results becomes more important in order to extract any subtle design flaws.

4. Once again, the three factors are:

- Movement (the user must be able to move naturally in the environment)
- Communication using human senses via stereoscopic vision or sound I/O systems
- Behavior that is complex and representative of the objects involved.

Such simulators fail the three criteria to differing extents.

The most obvious distinction between the flight simulator and an actual aircraft is the sensory inputs. The flight simulator does not load forces on the pilot, the control systems do not respond at different rates for a spectrum of airspeeds, there is no realistic dimming of the instrument panel due to bright sunlight, the sound of the engines and air flow is very unrepresentative, and many other factors.

The movement of the aircraft usually is unrepresentative. However, the simulator does provide motion and this defect may not be apparent to non-pilots. This motion is usually inadequate for high precision maneuvers such as landing, though, because the peripheral cues and depth perception are so poor.

Finally, the behavior of such simulators is much too predictable to be representative of actual flight. This deals not only with in-flight emergencies, but workload issues such as turbulence, communication congestion, requests for nonstandard procedures, and many other factors.

For these reasons, the FAA may verify simulators of this type as "procedure trainers," not simulators. Procedure trainers allow pilots to practice checklists and methodically perform Instrument Flight Rules (IFR) procedures. True simulators address the additional factors just outlined.

5. The temptation in building such systems is to provide lots of audio output and dazzling graphics. Many times, these happen so quickly that most viewers cannot fully comprehend the outputs. The thresholds are really human factors issues, but some rules of thumb are:

- Visual: >30 Hz for generation of continuous motion, <2 Hz for displays that require inputs
- Audio: outputs requiring a response should be widely separated in frequency and probably should not exceed six.

Exceptions to such rules relate to attempts to model physical phenomena whose rates of change are known. Sometimes the real world overwhelms us too.

Chapter seven

1(a). Usually, the performance requirements relate to the speed with which the system can run typical application software. For example, a personal computer is a single user system designed mostly for business applications such as word processing, financial spreadsheets, and preparation of presentation material. The performance requirements for such a system are likely to relate to delay times between user input and software response, support for the graphical user interface, and capability to load from high density media.

1(b). In the previous example, the interfaces are the keyboard for input and the graphics systems for output. Some applications might use other outputs such as a sound card or some other control interface. As a more concrete example, consider a flight simulator that might use a joystick or yoke controller for input in addition to the keyboard. The outputs are the "out the window" scene and sound effects.

1(c). Performance relates to how efficiently the system performs a function. Some functions for the flight simulator example might include:

- update "out the window" scene,
- update instrument panel,
- output sound file,
- accept control inputs,
- calculate new aircraft solution.

2. The system level requirements are likely to relate to the capability to cool the area in which the computer is installed. Consider the personal computer example again. A typical environmental restriction is the temperature range of 0 to 100°F. This often relates to the input power which is often in the range of 250 watts at around 80% efficiency. This means that the unit must dissipate at least 50 watts in heat, which provides some measure of the maximum temperature.

The system requirement can be partitioned down to module requirements. A worst case input signal might be 100 mA and there might be 256 signals at 5 volts with 20% loss, or 25.6 watts for the card. If there are 4 such cards, the system must dissipate $4 \times 25.6 + 50$ watts = 152.4 watts.

The final requirement relates to the cooling air. One usually assumes the worst case temperature (100°F) and a certain amount of air exchange in ft^3/ min due to the cooling fan.

This example shows how system power dissipation impacts the requirements for the power supply efficiency, digital module dissipations, and cooling fan design.

3. The installed volume for the personal computer example relates primarily to the typical dimensions found in an office. The system unit is likely to measure somewhere around $6'' \times 16'' \times 16''$ so as to allow the monitor, unit, and keyboard to fit on a standard sized desk. There is no major competitive advantage for decreasing the size beyond this, but there may be considerable cost differences. Therefore, there is little impetus to further miniaturize the design.

The situation is quite different for the laptop computers. The size requirements are often derived from the size of briefcases, so smaller computer sizes are necessary. Usually, the laptops are somewhat more expensive as a result.

4(a). A TPM formulation usually involves the current estimate of the performance requirement vs. program schedule. The earlier in the program, the more leeway is required to mitigate risk. However, overdesign is costly also, so extravagant performance is unwarranted.

4(b). The TPM should relate directly to the requirements. However, the purpose is to allow the program manager to look at "fuzzy" rankings such as high, medium, or low instead of needing to understand the engineering data directly.

4(c). Many program managers require lead engineers to track critical parameters, but unless the management requires such information, few engineers are likely to undertake such studies independently.

4(d). Currently, simulation is most often used for early phases of a program. The purpose of a VP capability is to extend this usage so as to serve as a bridge between automated design and automated manufacturing.

5. Revisit the answers to the last question.

6. The discussion should address how the VP capability should emulate the actual processes involved in engineering and manufacturing. As usual, only the critical elements require emphasis in the process model. Such elements are primarily those that relate to technical risk, cost, and schedule.

7(a). Requirements specification suffers the same imprecision as many technical communications in a spoken language. Mathematical models can describe some events with precision. The use of an artificial requirements language may solve such problems. This might be an existing software language such as Ada or VHDL so that an automated method exists to check requirements syntax and grammar.

7(b). It may be possible to use high resolution graphics to provide a more detailed glimpse at the requirements (literally!) instead of the language interface. Discuss some visual presentations. Again, for initial designs, a qualitative ranking may be sufficient. Does fuzzy logic help?

7(c). Interface standards allow the tools to be more modular. This provides for incremental upgrades of the capability as a particular tool becomes obsolete. The drawback might be performance as point designs usually are faster than standard approaches.

7(d). The advantage of a language such as VHDL is the wide industry support which offers the user mature, relatively low cost tools. If one uses simulation models to extract achievable performance, one still must build some kind of post processor to compare the results to requirements.

Chapter eight

1. The most common comparisons are between C++ and Ada. C++ offers specific OOD support through facilities such as the class structure. Ada does not provide this facility but offers the much clumsier "generic" that can be used in a similar manner to define objects. The inheritance mechanism, for example, is much cleaner in C++ but overzealous usage can produce complex behavior (similar to "spaghetti code"). Both languages support operator and function overloading and polymorphism.

2(a). Instruction sets differ among designs because of the numerous options in the internal architectures and external memory systems. However, it is possible to build an instruction translator that can run in real time. These are similar to the instruction set simulators used in software development. The performance penalty for such implementations is normally quite high, however.

2(b). A common assembly language offers many of the same benefits as a common Higher Order Language. The drawback is that a true assembly language standard demands a common instruction set with all of the drawbacks discussed in the previous answer. However, a common assembly language might be more of a functional standard. Many early Ada compilers, for example, converted Ada source code into C language source. The distinction is that C is less structured and formal and therefore functions more like a machine language.

3(a). Automata theory is the representation of computing machines as state machines and models of languages with which to program such state machines. This provides a foundation for both functional and interface requirements. In some cases, the models are also usable in performance analysis.

3(b). Queuing theory deals with the statistical behavior of queues. A queue is simply a waiting line for access to some type of service. This theory is quite useful in the performance analysis of computer networks assuming an on-demand service request model.

3(c). Scheduling theory involves ordering and timing of access to a service. In operating systems, scheduling is used to execute software processes and to mediate the access of such processes to hardware resources.

3(d). An analyst must often use several different types of models to perform a thorough investigation of system behavior and performance. The key is to select the proper model types based on the intended usage of the customer. It might be inappropriate, for example, to use models from queuing theory for a control system whose software processes are highly repetitive and predictable, not statistical.

4(a). This is an issue of fidelity vs. complexity. A gate level description might allow the designer insight into detailed electrical and timing characteristics but such a simulation is likely to be so large that it will be very expensive to debug and execute slowly. The other consideration is risk. In the design of a memory interface for a processor, small arrays of memory often give nearly as much hardware design insight as in large arrays.

4(b). It is possible to disassemble compiled code and then sum up all of the clock cycles. More specifically, suppose that instruction A requires 11 clocks and instruction B requires 17 clocks. If the disassembled code contains 100 occurrences of A and 200 occurrences of B, then (in theory) the code requires 4500 clock cycles. If the processor clock speed is 12.5 MHz (80 ns) then the mix requires 36 μs. In more familiar terms, the mix contains 300 instructions within the 36 μs, which yields to an average of 8.3 millions of instructions per second (MIPS).

There are numerous flaws in this analysis. First, the instruction sequence, not just the mix, determines the number of clock cycles for execution. For example, a string of A instructions in sequence may require 11 clocks, but because of resource contention, a sequence of interleaved A and B instructions may result in much different results. Also, this simple analysis also does not account for other resource contentions such as collisions on the memory bus interface.

4(c). It is quite possible that much of the existing simulation is reusable. The memory arrays are the obvious example, but any interface logic may be partially reusable from the memory side. The processor side of interface logic likely will require redesign.

5(a). To simulate the end to end latency, one needs simulations of all of the elements. This includes the data bus, but also memory and processor systems.

5(b). Many design tools accept STDs and timing diagrams as inputs. If these diagrams are available, it may be possible to automatically generate simulation code.

5(c). If the basis of the electrical design is some standard, then it may be possible to select such elements as part of a tool to automatically generate simulation code. An example is transistor-transistor logic (TTL).

The electrical design often is the performance limitation, not necessarily the bus protocol. Hardware components such as transceivers operate at fixed upper rates that depend only upon the silicon process used in their manufacture. A common mistake is to disregard such effects and predict unattainable performance levels based solely on the operation of the bus protocol.

5(d). The STD may show the bus error conditions. If so, the code generator may incorporate these directly. However, the message scenario must also include such events, not simply normal operation.

Chapter nine

1(a). Typical processor boards available are configured with a data processor and 1 to 4 MB of program memory. The module size usually depends on the parallel bus interface type, but the Eurocard 6Ux150 is very common.

1(b)–1(e). Revisits the design rules. Simply pick convenient requirements to make the calculations tractable.

2(a). Revisit the CIPO model. At the highest level, such applications usually involve three top level functions:

- Sensor interface operation
- Control processing
- Control outputs.

The sensor interface operation might involve units such as sensor updates (transfer the sensor values to the processor periodically) and error checking (verify that the control inputs and sensor values are not diverging). There are likely to be support routines such as a trigonometric math library.

To provide an estimate for lines of code, simply multiply the number of units by 250 (average unit size instead of maximum possible). In this case, the transfer and error checks might involve 2 units each for a total of 1000 lines of code.

2(b). The 1000 lines of code represents 4000 instructions. A 20 MIPS machine could execute this code in:

$$4000 \text{ instructions}/[20 \times 10^6 \text{ instructions/s}] = 2 \times 10^{-4} \text{ s.}$$

Usually this is much faster than the sensor update rate. If the rate is periodic at 100 Hz, this portion of the function only uses 2% of the utilization.

2(c). 4000 instructions × 4 bytes/instruction = 4000 bytes. To convert this to KB, divide by 1024 to get 3.9 KB. The total storage (code + data) is twice this value or 7.8 KB. For 1 MB available storage, this represents 0.78% utilization.

2(d). For classes of problems, the relationship between code size and performance requirement might be modeled linearly just as was the relationship between I/O and performance. This relationship appears to be much less general, though, as the coefficients of the equation appear to change among applications.

3(a), (b). These will depend on the hardware and software selected.

3(c). The development cost is (assume $100 per hour for round figures):

24 Man Months × 320 hours/MM × $100/hour = $768,000.

The maintenance cost is:

2 persons, 5 years = 60 Man Months = $1,920,000.

3(d). The design cost of the automated system is $750,000. The cost of a typical application, 3(c), is $768,000. If the automated design requires only 1/10 this cost, then the savings per design is 9/10 of $768,000 or $691,000. This means that the break even point for the automated system is two designs.

4(a). This question deals with scaleability. It will not be possible to simply add processor modules indefinitely. The obvious factor is that the enclosure will run out of growth slots. However, performance is also a factor. In general, performance does not scale linearly. In fact, usually after about four to five processor modules, it is more efficient to move to a two computer system.

4(b). The large increase occurs at the end game when one must convert a single computer into a network. Of course, technology helps somewhat here because more powerful cards are available each year. Nonetheless, at some point, a network is more efficient in terms of cost because many resources are not used by a single processor.

4(c). If only a single type of module is used, the addition of new circuit cards decreases the reliability in the ratio of circuit card counts:

reliability = (old reliability) × (old module count/new module count)

Once again, the computer network involves a decision point because the reliability of the system is likely to go down significantly at this point.

4(d). This revisits the issue of delaying upgrades by waiting on technology. Ordinarily, the new technology boards are more expensive, so one can reach this break even point quickly.

5(a). A graphical interface might be most intuitive. The graphical problem solving method should not be an arcane design language, but should illustrate the key features of each technology element.

5(b). A multimedia interface might be efficient. This could consist of high resolution graphics read from mass storage, sound prompts, or even voice inputs. The stereoscopic goggles might be useful for considering the installation of the computer equipment in its operating location.

5(c). With some design effort up front, an OOD approach might allow much of the simulation code to be reused. The interface code is an example. If possible, use a common graphical interface for all three stages so that the widest possible community can use all three tools.

Chapter ten

1(a). 78 clocks per message are needed for a total of 780 clocks. At 20 MHz, each clock is 50 ns, so the bus operations require 39 µs.

1(b). The software latency is a function of the number of data words and control words. If the host side memory bus is also 20 MHz, this total time is the same as the bus operation time just calculated. However, the software scheduling delay is based on when the process executes. The schedule table is:

Process	Ticks	Total ticks
1	2	2
10	2	4
2	4	8
9	4	12
3	6	18
8	6	22
4	8	30
7	8	38
5	10	48
6	10	58

If each tick value is 0.1 ms, the total scheduling delay is 5.8 ms.

1(c). If function 10 performs all of the I/O, the total delay is software delay + hardware delay = 5.8 ms + 0.39 ms = 6.19 ms.

1(d). A total of 4 procedure calls in this example requires 600 µs. An interrupt response usually is much faster than a procedure call (e.g., 10 µs). This demonstrates that software abstraction can extract a performance penalty. However, most modern processors have enough performance so that this is not a significant concern.

1(e). The calculation is the same as 1(d) except that the order is reversed.

1(f). The total time, 1(a)–1(e), is $2 \times 5.8 + 0.39 = 11.99$ ms.

2. Consult Ref. 2 of this chapter.

3, 4. Refer to this chapter for examples.

5(a). As discussed in the text, the design and VP engines might be the same software. For VP in particular, one must have the bus and module libraries to use as building blocks. In addition, one must have some type of test scenario.

5(b). The initialization consists primarily of the message tables for the trade study. This is the source, destination, and time to schedule each message.

5(c). This relates back to the survey of work stations and AI tools as a project from Chapter 9.

5(d). The message mix is one aspect of the test scenario for performance. Another is some type of input that presents a threshold for the minimally acceptable throughput. The test scenario might also perform a sensitivity analysis in which various parameters might vary. For example, one consideration might be the variation with respect to MIPS that a module provides.

Acronyms

ADM	Advanced Development Model
AI	Artificial Intelligence
ALU	Arithmetic Logic Unit
ARTCC	Air Route Traffic Control Center
ASIC	Application Specific Integrated Circuit
CASE	Computer Aided Software Engineering
CIPO	Control, Input, Process, Output
CMM	Capability Maturity Model
CMOS	Complementary Metal Oxide Semiconductor
COEA	Cost and Operational Effectiveness Analysis
CSSR	Cost Status and Schedule Reporting
DoD	Department of Defense
DoDISS	DoD Index of Standards and Specifications
EDM	Engineering Development Model
FC	Flow Control
FPU	Floating Point Unit
GLU	General Logic Unit
GOMAC	Government Microcircuits Applications Conference
GPS	Global Positioning System
HDL	Hardware Design Language
HOL	Higher Order Language
IBIT	Initiated Built in Test
IDE	Interactive Development Environments
IEEE	Institute of Electrical and Electronic Engineers
IFS	Iterated Function System
IPDT	Integrated Product Development Team
IPL	Intermediate Pass Languages
ISA	Instruction Set Architecture
LORAN	Long Range Area Navigation
MCM	Multiple Chip Module
MMU	Memory Management Unit
MTBF	Mean Time Between Failures
MTTR	Mean Time To Repair
NAECON	National Aerospace and Electronics Conference
NAS	National Airspace System
NCOSE	National Council on Systems Engineering
NL	Natural Language
OOA	Object Oriented Analysis
OOD	Object Oriented Design

OMT	Object Modeling Technology
OOP	Object Oriented Programming
OOSA	Object Oriented System Analysis
OOSD	Object Oriented System Design
PBIT	Periodic Built in Test
PWB	Printed Wiring Board
RTM	Requirements Traceability Model
RTL	Register Transfer Language
RTS	Run Time System
SA	Structured Analysis
SAE	Society of Automotive Engineers
SD	Structured Design
SEI	Software Engineering Institute
SQL	Standard Query Language
STD	State Transition Diagram
StP	Software through Pictures
TACAN	Tactical Area Navigation
TPM	Technical Performance Measures
TQM	Total Quality Management
TTL	Transistor-Transistor Logic
VHSIC	Very High Speed Integrated Circuits
VHDL	VHSIC HDL
VP	Virtual Prototyping
VR	Voice Recognition

Abbreviations

μs	Microsecond, 10^{-6} seconds
Hz	Hertz, units of seconds^{-1}
KB	Kilobytes, 1024 bytes
ms	Millesecond, 10^{-3} seconds
MB	Megabytes, 1024 KB
ns	nanosecond, 10^{-9} seconds

Index

A

Abstraction, 48, 132
Abstract model building, 224
Access type, 149
Acronyms, 253–254
Ada, 42
  code, 149
  compilers, 248
  fragment, 48
  HOL, 148
  implementation, 49
  source code, 248
Adder,
  multiple bit, 79
  single bit, 77, 78, 242
  three input, 77
Adequate capacity signal, 139
ADM, see Advanced development model
Advanced development model (ADM), 108,
  109
AI, see Artificial intelligence
Aircraft
  avionic systems, 139
  diagnostics, 99
  embedded computers, 14
  heading, 104
  systems control, 96
Air route traffic control centers (ARTCCs),
  135–137
Airspace system, model of, 135
Algorithm example, 167
Alpha release, 109
ALU, see Arithmetic logic unit
Analog instruments, 87
Analysis
  distinction between design and, 35
  methods, 3
  rules, 163
Apollo effort, 9
Application software, 192
Architecture
  block diagram, 175

design, 94
elements, 172
notion of, 78
reengineering, 186
topology, 169
Arithmetic logic unit (ALU), 78
ARTCCs, see Air route traffic control
  centers
Artificial intelligence (AI), 2, 93
Assignment
  algorithm, 166
  criterion, 183
Associative memory, 104
Automata theory, 248
Automated design process, 195
Automated teller machines, 12
Automated tools, 58, 195
Automatic document generators, 213
Avionic
  hotbench, 87
  weights, 125

B

Batch job, 133
Behavioral simulation, 80
Beta testing, 108, 195
Black box, 41, 51, 235
Block diagram, 174
Blocking
  messages, 169
  send, 205
Blueprint redlining, 95
Booch notation, 60
Brassboard unit, 108
Brownian motion, 31
Burst rate, 172, 173
Bus
  list, 202
  side model, 200
  simulations, 154